Information and Communications
in the Chinese Countryside

A WORLD BANK STUDY

Information and Communications in the Chinese Countryside

A Study of Three Provinces

THE WORLD BANK
Washington, D.C.

Contents

Boxes

Figures

Map

Tables

Foreword

China is set to become one of the East Asia and Pacific region's most "connected" countries and plans to upgrade its connectivity radically over the next few years. The government's Broadband China Strategy, approved by the State Council in August, 2013, characterizes high-speed or broadband Internet as strategic public infrastructure for the economic and social development of China in a new era. Investments in broadband Internet are intended to stimulate new opportunities through connectivity, particularly in more rural and remote areas of the country in order to reduce the "digital divide." Key targets include provision of broadband Internet access for 98 percent of China's villages, and Internet speeds of 12 megabytes per second by 2020 for rural householders.

Information and communications technologies (ICT) and associated applications have potentially transformative potential for China's countryside. ICT applications may support, among other things, improved rural productivity, by facilitating access to agricultural information, services and markets; e-commerce and financial inclusion; supply of teaching and learning materials for younger generations; income-generating and business development opportunities beyond traditional agriculture; and communications, skills development, and improved social inclusion for senior citizens.

This report considers, from the grassroots level, the impact from investments in rural ICT in selected provinces, and what can be learned from these experiences. The report seeks to understand the key drivers of demand for ICT services from the perspective of the rural population and highlight some innovative approaches taken to use ICT for agricultural productivity enhancement. A notable feature of the report is an assessment of the role of public libraries whose network extends from major metropolitan centers all the way to the village level. The main messages emerging from the report acknowledge the significant progress in improving basic access to ICT infrastructure, and emphasize the need for a stronger focus on demand-side interventions, user skills development and outreach, and institutional collaboration.

The World Bank appreciates the cooperation with a number of national and provincial institutions in the preparation of this report, in particular the National

Development and Reform Commission's High-Tech Department, and looks forward to future collaboration in realizing the cross-cutting benefits of ICT for the Chinese countryside.

Klaus Rohland
Country Director
China, Mongolia, and Korea
East Asia and the Pacific Region
The World Bank

Acknowledgments

This report is based on research funded in part by the Bill & Melinda Gates Foundation. The findings and conclusions contained within are those of the authors and do not necessarily reflect positions or policies of the foundation or the World Bank.

The publication was written by Michael Minges, Kaoru Kimura, Natasha Beschorner, Robert Davies, and Guangqin Zhang. Overall guidance for the publication was provided by Klaus Rohland (Country Director, China), Mark Lundell (SD Sector Manager, China), and Randeep Sudan (Sector Manager, ICT Unit). Valuable contribution and comments were provided by Junko Narimatsu, Lihong Wang, Garo Batmanian, Jun Zhao, Elena Glinskaya, Tim Kelly, Aparajita Goyal, Christine Zhenwei Qiang, Ana Goicoechea, Jing Cai, and Li Li (World Bank).

The authors are grateful to the High-Tech Department of the National Development and Reform Commission of China and State Information Center for their support.

The publication was edited by Colin Blackman.

About the Authors

Michael Minges is an independent consultant with more than 30 years of experience in the area of ICT for development. He regularly advises governments on key ICT policy, strategy, and regulatory issues and has written widely about telecommunications progress in developing countries. Prior to becoming a consultant, he led the market and economic analysis team at the International Telecommunication Union.

Kaoru Kimura is an ICT policy specialist with the ICT Sector Unit at the World Bank Group. She has worked on several operational and analytical projects in Sub-Saharan Africa and East Asia. She has been working for ICT policy issues in China since 2006, and she is a coauthor of "Rural Informatization in China" (World Bank Working Paper No. 172, World Bank 2009). Before joining the ICT Sector Unit, she worked at Nippon Telegraph and Telecommunication in Japan.

Natasha Beschorner is a senior ICT policy specialist and regional coordinator for the World Bank's ICT engagements in East Asia & Pacific. She has been with the World Bank since 1993. Her main responsibilities include design and management of investment and policy-based projects and technical assistance programs on ICT for development in China, Indonesia, the Philippines, Papua New Guinea, Timor-Leste, and the Pacific Islands.

Robert Davies is a chief executive of MDR Partners, a consulting firm based in London, United Kingdom, working in the area of libraries and information and specializing in strategies for their development in the digital era. He has more than 20 years of experience in studies for and preparation of donor-funded initiatives in various countries, including South Africa, Lao PDR, the Arab Republic of Egypt, Jordan, Namibia, and Greece.

Guangqin Zhang is an associate professor at Peking University, Department of Information Management, in China. He has been presiding over some research projects from National Philosophy and Social Science Fund Committee, Ministry of Education, China Postdoctoral Science Foundation and other types of organizations. His research interest includes public library: service, evaluation, and building standard; Chinese library science and librarianship; and information resource management.

Abbreviations

ADSL	Asymmetric Digital Subscriber Line
BTT	Business and Technology Telecenter
CBC	Community Broadband Centers
CCSCP	Comprehensive Cultural Station Construction Project
CCTV	closed-circuit TV
CD	compact disc
CIRSP	Cultural Information Resource Sharing Project
CMMR	China Mainland Market Research Co., Ltd
CNNIC	China Internet Network Information Center
CNY	Yuan Renminbi, the unit of China's official currency. Conversions to U.S. dollars have been made on the basis of the 2012 annual average exchange rate (CNY 6.3093 = $1)[1]
CPC	Communist Party of China
GAEN	Guizhou Agriculture and Economy Network
DLPP	Digital Library Promotion Project
DVD	digital versatile disc
ICT	information and communications technologies
IE	Information Express
IT	information technology
MCMC	Malaysian Communications and Multimedia Commission
MIIT	Ministry of Industry and Information Technology
MRI	Mountainous Region Informatization
MT	multipurpose telecenter
NGO	nongovernmental organization
NLC	National Library of China
OLPC	One Laptop per Child
PC	personal computer
PPP	public-private partnership
REAP	Rural Education Action Program

REIN   Rural Economic Information Network
RFID   radio-frequency identification
SIC   State Information Center
SMEs   small and medium enterprises
SMS   Short Message Service
UISC   Union Information Service Centers
USDA   United States Department of Agriculture
VCD   video compact disc
VCR   video cassette recording

All dollar amounts are U.S. dollars unless otherwise indicated.

Note

1. Board of Governors of the Federal Reserve System. 2013. "Foreign Exchange Rates: G.5A Annual." January 2. http://www.federalreserve.gov/releases/g5a/current/.

Executive Summary

Improving access to information and communications technology (ICT) and related services in the countryside—or rural *informatization*—is a long-standing Chinese policy objective. National and provincial governments and China's ICT industry have invested significantly in rural infrastructure and facilities over the past decade with the goal of reducing the country's digital divide. The purpose of this study, undertaken at the request of the Chinese government, is to review this experience and inform future approaches to rural informatization.

The study focuses on three provinces with different socioeconomic characteristics: Shandong, Jilin, and Guizhou. The scope of the study included (a) a demand survey to assess rural ICT access and usage; (b) a review of ICT in primary and secondary schools; (c) a survey of public libraries, including the extent of ICT use in rural libraries; and (d) an assessment of specific ICT interventions to examine how they have affected rural users. Much of the published information about rural ICT development in China describes infrastructure deployment, with top-level target monitoring statistics. This report sheds light on findings at the grassroots level through surveys and interviews, exploring the nature of demand for ICT services from rural populations, and considers whether this demand is being adequately addressed. Though there are differences in infrastructure and access across the three provinces, the structural challenges are similar. The lessons learned are not only consistent across the three provinces but also similar to research findings on rural informatization in other provinces. Thus, they are likely to be relevant for making recommendations about future approaches in other rural areas in China.

Notwithstanding the steady urbanization trend in the country, the welfare of rural citizens remains an important issue in China. There are over 650 million people residing in the country's rural areas, the second largest rural population in the world after India. Further, China is witnessing a demographic shift in villages as many young adults move to urban areas in search of work, leaving behind older people and children. ICTs can enhance the lives of rural citizens by helping them to stay in touch with family members; provide information about agricultural and off-farm livelihoods; deliver public services; and disseminate educational and health content. In order to leverage these benefits, rural residents need to have access to ICTs and know how to use them. Despite significant rural informatization investments by both the government and private sectors, the gap in ICT access between urban and rural areas in China is noteworthy. In 2012, Internet penetration in urban areas was more than two-and-half times the level in rural areas.

Access to ICT infrastructure and services is improving at the village level. The rural ICT demand survey noted that all villages had electricity and received a mobile telephone signal. Ninety-seven percent of households owned a television and 95 percent had access to a mobile phone. However, only about 20 percent of surveyed households owned a personal computer or used the Internet, which they attributed primarily to lack of skills and awareness of the benefits. ICT training opportunities were limited, however. While many rural schools had Internet access, only a quarter reported making their facilities available to the local community and less than 10 percent of public ICT facilities reported that they offered training. Although not the main bottleneck, affordability remains a barrier for some, particularly for the purchase of computers, tablets, and smartphones.

Most rural Internet users used Internet café facilities in townships. Four out of five of these were privately operated, and the vast majority of users were young men. Interviews with the managers of Internet cafés indicated that the top three activities performed by users were online gaming, social networking, and watching videos or downloading songs. While provincial authorities have devoted considerable resources to developing agricultural websites, less than 2 percent of all rural household respondents reported using the Internet to access online government agricultural information.

There is a noticeable urban/rural digital divide in schools. Government programs aim to reduce this gap but survey results highlight the continued lower level of ICT equipment and usage in rural primary schools. On the other hand, it is notable that about 95 percent of those aged 10–19 were online in China in 2012.[1]

The Chinese public library system offers opportunities to bridge the digital gap. With their substantial community-facing infrastructure, libraries could become effective instruments of delivery in the provision of digital literacy and ICT skills development. Several national projects have been established to provide library and information services in rural areas. Their effectiveness is restricted by a number of factors including insufficient operational funding beyond the investment stage; challenges with recruiting and paying staff adequately trained in service management and delivery; limited ability to keep facilities open; uneven network connectivity and equipment maintenance problems. The absence of effective administrative networks for libraries, which could enable larger libraries in urban areas to support and act as hubs for rural libraries in a region, is a further impediment to effective delivery. Despite these constraints, the data obtained in this study suggest a more intensive use in rural areas than in more mainstream urban library settings, most likely because there are fewer time-consuming alternatives.

Strengthening the evaluation of rural ICT initiatives could improve results in the longer term. Several rural ICT projects were reviewed in order to assess their impacts. The interventions have benefited those who use them in various ways. Well over half the users reported monetary benefits from productivity gains due to learning better production techniques, buying agricultural inputs at a cheaper price, and selling farm products through new sales channels. Nonagricultural benefits include acquiring Internet-surfing skills, thereby learning more about the outside world and broadening one's horizon. However, these interventions currently focus primarily on (male) farmers. Nonfarm workers, the unemployed, women, and senior citizens hardly use these facilities and services, citing limited skills or perceived relevance. While the emphasis on agriculture reflects current economic priorities, a more forward-looking approach could also develop applications and programs that support skills development and ICT-enabled contract work that might appeal to a broader user group. The type of intervention was found to have an impact on use. Text messaging, listening to the radio, or watching TV are largely passive. In comparison, a personal visit to a telecenter requires a certain degree of initiative; calling a hotline costs money; and visiting a website requires access to computers and the Internet as well as information technology (IT)-related knowledge.

There has been significant progress in extending ICT infrastructure to rural areas of China. By the end of 2012, all of the nation's administrative villages had been connected to the telephone network, broadband connectivity was available in 88 percent of administrative villages, and there were 156 million rural Internet users. Extensive agricultural content has been created through national and provincial initiatives. Results from the program's study of impacts also reveal positive benefits for rural inhabitants who use ICT facilities and services. Nonetheless, there remain sharp differences in Internet use between urban and rural areas in China. Further, most rural interventions are generalized, primarily agricultural applications aimed at enhancing farmer welfare with few developed for those with off-farm livelihoods.

However there are significant opportunities for improvement. Building on the experience of rural ICT initiatives implemented to date and incorporating international best practice, the government should consider new models for rural informatization to achieve wider inclusion and higher impact with economies of scale and lower investment costs. Recommendations include the following:

- **A more coordinated approach**. Similar telecenter, e-commerce, and mobile phone information systems developed by each province result in duplication of resources. This could be remedied through improved coordination by the central government with the aim of scaling up and standardizing common applications and services. In that respect, the government might consider creating a consultative committee consisting of key ministries involved in rural informatization (for example, Industry and Information Technology, Agriculture, Culture, Education, Science and Technology, Commerce, Finance, and so on) to be responsible for overall coordination.

- **Demand stimulation.** Demand for ICT services in rural areas could be increased through more inclusive applications and services to enrich the quality of life all rural citizens. Computer training needs to be expanded particularly for older people and women. Ongoing outreach campaigns such as open days, competitions, and marketing at local gatherings should be developed to raise awareness of rural ICT programs. Successful impacts should also be demonstrated so that villagers can see the concrete benefits of ICT use.
- **Improved monitoring and evaluation.** One of the reasons for duplicate systems across provinces is the shortage of information regarding the outcomes of various programs. Data on the results of rural ICT interventions should be collected, compiled, analyzed, and disseminated on a regular basis, and should be used for future policy decision making. Users and nonusers of the various programs should be surveyed both ex ante and ex post to generate credible evidence regarding impacts.
- **Stimulating innovation.** Public Internet access facilities such as telecenters could be leveraged to become hubs of village innovation through networking rural businesses and communities. This includes developing linkages with international and domestic firms for ICT-enabled contract work. The use of e-commerce and social networking platforms by small business should be fostered in order to improve productivity and expand access to domestic and global markets.
- **Focus on sustainability.** There are several practices to achieve greater long-term sustainability for rural ICT interventions. This includes making greater use of partnerships and shifting the obligations of telecommunications operators from infrastructure deployment to operations. Libraries, in particular, are logical partners for rural informatization with their community-facing infrastructure. Public-private partnerships should be more broadly utilized for content and applications to leverage the expertise of China's Internet companies. Income-generating services could be incorporated in public Internet access facilities to defray costs and increase demand.
- **Complementarity of access devices and facilities.** ICT access from computers and mobile devices should complement each other. Mobile phones are ideal for personalized, short sessions, and small information streams. Computers and tablets are suited for more intensive applications, extensive searching, and e-book reading. Telecenters should extend their portfolio by more deeply integrating wireless support. At the same time, considering the likely growth of smartphones and other Wi-Fi-enabled mobile devices, the government could consider supporting free wireless local area network access for rural areas.

Summary recommendations are provided below (see table ES.1). These holistic approaches, featuring delivery of appropriate information and entrepreneurial management and supported by clear measurement results, are likely to have wide economic benefits for rural communities, with the goal of integrating Chinese villages into modern society.

Table ES.1 Policy Matrix of Recommendations

Recommendation	Beneficiaries	Actions	Institution	International example
A more coordinated approach	Program implementers	Coordinate rural informatization activities; integrate similar programs to minimize duplication; disseminate information about best practice; provide information about central government programs and support; audit deployment and enforce operating standards	Steering committee of government institutions involved with rural ICT development	Australia *Digital Regions Initiative*
Demand stimulation	Seniors	Target training and applications	Central/Provincial government	Australia *Broadband for Seniors* program
	Women			Philippines *Digital Literacy for Women Campaign*
	Children			Uruguay *Plan Ceibal*
Monitoring and evaluation	Program implementers	Ongoing reports on usage or rural ICT interventions and assessments of their impact	Provincial government	Latvia and Lithuania outcome measuring reports on impact of public ICTs
Innovation	Local communities (SMEs)	Leverage telecenters and libraries to become hubs of village innovation	Local government	U.S. *Business and Technology Telecenters*
Sustainability	Local communities	Universal service fund for rural ICT activity operation	MIIT	Malaysia *Universal Service Provision* program
	Local communities	Leverage libraries for public ICT access	Ministry of Culture	Latvia *Trešais tēva dēls* ("Father's third son")
Mobile and computer synergy	Local communities	Provide Wi-Fi; Develop localized apps; Offer printing/scanning services	Provincial authorities and Internet cafés	Thailand *Free Wi-Fi*

Source: World Bank.
Note: See chapter 7 for additional details about international examples. ICT = information and communications technology; MIIT = Ministry of Industry and Information Technology; SMEs = small and medium enterprises.

Note

1. Derived from China Bureau of Statistics and China Internet Network Information Center (CNNIC) data.

Information and Communications in the Chinese Countryside
http://dx.doi.org/10.1596/978-1-4648-0204-1

CHAPTER 1

Introduction

In its last few development plans, the government of China has emphasized the importance of harmonious development across all regions of the country, including remedying divides between urban and rural areas. This includes expanding access to information and communications technology (ICT) in rural areas, or what the Chinese government calls rural *informatization*.

The World Bank has been supporting this policy through a number of activities including "China Rural Information and Communications: Technical Assistance on Design and Impact Evaluation." This analytical/advisory program aims to support the government to make decisions about potential scaling up of innovative ICT pilot projects and to generate and disseminate knowledge about the impacts of ICT in rural China.

In light of these objectives, three activities have been undertaken to date: (a) a demand survey to assess rural ICT access and attitudes; (b) a library study including scoping the status of ICT use in rural libraries; and (c) a limited impact evaluation to examine how ICT interventions have affected rural users.

The approach, experiences, and outputs of the three activities offer a distinct perspective on rural informatization. Much of the information available about rural ICT development in China is focused on infrastructure deployment and typically reinforced by top-down administrative statistics. This report sheds light on findings at the grassroot level through surveys and interviews. The activities explore the nature of demand for ICT services from rural populations and analyzes whether these are being adequately addressed. Although the program focused on a trio of provinces (Guizhou, Jilin, and Shandong), many of the lessons learned were consistent across the three and thus are likely to be relevant for making recommendations about future approaches in other rural areas in China.

Chapter 2 provides background information on China's rural informatization policies and status. The demand survey is discussed in chapter 3. ICT use in rural schools is discussed in chapter 4. China's library landscape is outlined in chapter 5. The result of the impact assessment is highlighted in chapter 6. The last chapter offers conclusions and recommendations for advancing rural informatization in China.

CHAPTER 2

Interventions to Reduce
the Urban-Rural Digital Divide

Notwithstanding the steady urbanization trend in the country, the welfare of rural citizens remains an important issue in China. There are over 650 million people residing in China's rural areas, the second largest rural population in the world after India. Further, China is witnessing a demographic shift in villages as the young move to urban areas in search of work. This is resulting in a growing proportion of minors, women, and especially older people remaining behind in rural areas.[1]

In its 11th Five-Year Plan (2006–10), China adopted a new development paradigm that emphasizes the building of a "new socialist countryside,"[2] and a "harmonious society" with more balanced development across regions and across sectors. The new paradigm adopts a view of the development process that emphasizes sustainable growth and "putting people first." Under this paradigm, the government has substantially increased its commitment to pro-poor, pro-rural programs. This is a timely shift in policy to redress the large disparities that have emerged across sectors and regions in the course of China's economic growth.

A commitment to rural areas continues with the 12th Five-Year plan (2011–15) (United States Department of Agriculture [USDA] Foreign Agricultural Services 2011). It calls for accelerating the "development of the socialist new countryside" and, while emphasizing support for farmers, also seeks to encourage nonagricultural employment including the establishment of new businesses and jobs.

This chapter reviews actions by the Chinese government to reduce the digital divide between urban and rural areas. The central government has developed a number of policies, strategies, and programs for rural informatization. A notable feature of the Chinese information and communications technology (ICT) policy environment is that, while the central government typically sets out general guidelines and commonly funds a significant portion of the projects, design and implementation is carried out by provincial governments.

Central Government Rural Informatization Initiatives

Given that it is the world's most populated nation, it is no surprise that China has the largest telecommunication networks of any country. This includes the most fixed-telephone, mobile-telephone, fixed-broadband, and mobile-broadband subscriptions as well as the highest number of Internet users (see table 2.1). On a per capita basis, China is still some distance behind developed economies in most ICT penetration ratios. However, it is quickly narrowing the difference. In 2013, the State Council released the country's *National Broadband Plan*. It sets forth a number of coverage, penetration, and speed targets to be achieved by 2015 and 2020. Meanwhile China currently exceeds the four targets to be achieved by 2005, established by the Broadband Commission relating to affordability, penetration, and access (Broadband Commission 2013).

Table 2.1 China's ICT Statistics, 2012

		Per 100 people	
	Total (millions)	China	High-income economies (2011)
Fixed-telephone subscriptions	278	21	45
Mobile subscriptions	1,112	82[a]	122
Fixed-broadband subscriptions	177 (1/13)	13	25
Mobile-broadband subscriptions	233	17	—
Internet users	564	52% of age 6+	75

Source: China Bureau of Statistics and World Bank.
Note: — = Not available.
a. Unlike many other countries, the incidence of dual SIM ownership in China is low. This tends to underestimate the penetration rate compared to other nations. According to a 2012 survey, 93 percent of the adult population used a mobile phone (see Pew Research Center 2013).

The gap in ICT access between Chinese urban and rural areas is notable. In 2012, Internet penetration in urban areas was two-and-a-half times the level in rural areas (see figure 2.1). Moreover, the gap is growing with the difference in Internet penetration rates between urban and rural users rising from 14 in 2005 to 36 in 2012.

Figure 2.1 Internet Users per 100 People in China, 2005–12

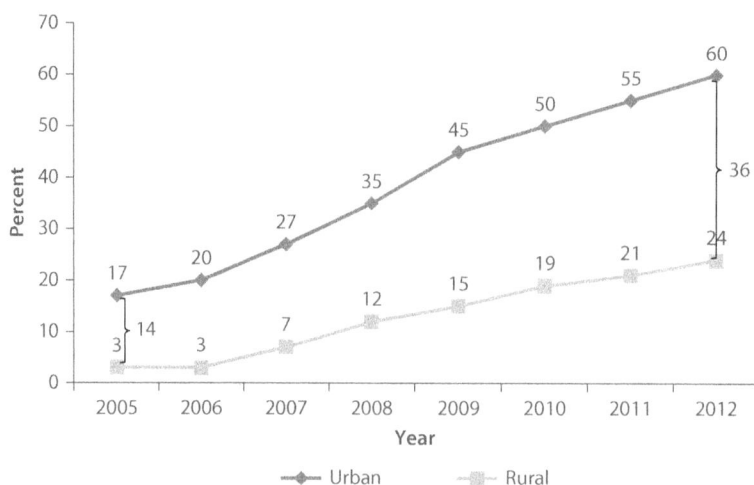

Source: Adapted from CNNIC.
Note: CNNIC = China Internet Network Information Center.

The central government has initiated a number of policies and projects to reduce the rural-urban informatization divide. Various ministries have introduced different interventions including infrastructure deployment, content development, and distance education (see appendix A). Key initiatives include (Yongtao 2009):

- **Villages Access Project** (also referred to as "Village-to-Village" and "Village Connected Project"). Launched in 2004 and coordinated by the Ministry of Industry and Information Technology, this scheme compensates telecommunication operators for extending telephone service to villages. By 2010, all administrative villages and 94 percent of natural villages had access to the telephone network (see figure 2.2, a). If mobile coverage is factored in, then telephone coverage is almost ubiquitous.[3]
- **Home Appliance Subsidy Program** for rural areas. In 2009, computers were included in the rural home appliance subsidy program. The subsidy provided a discount of 13 percent for those living in rural areas for the purchase of a home appliance approved by the Ministry of Commerce (He and Ye 2009). The program ended in January 2013. While the number of computers per 100 rural households rose to 18 in 2011, up from 2.1 in 2005, there is no formal evidence attributing this to the appliance subsidy program (see figure 2.2, b). The increase of computers in rural households grew faster than any other appliance in 2011; nonetheless, the penetration of computers remains the lowest of any major consumer durable.
- **Information to the countryside**. Launched in April 2009, this multifaceted activity involves developing content, services, and access points for rural areas. One of the guidelines for public access is the so-called "Five One," meaning a single place where access is provided, one set of computers, a responsible person (that is, "information servant" or "information worker"), one set of rules, and a mechanism for sustainability. It is also recognized that information delivery could be via different platforms (broadcasting, telephone and text messaging hotlines, and websites). Telecommunication operators have been providing service platforms for delivering the content as a part of their corporate social responsibility programs (see box 2.1).

Figure 2.2 Percentage of Villages with Telephone Services and Computers per 100 Rural Households

a. Percentage of villages with access to telephone network

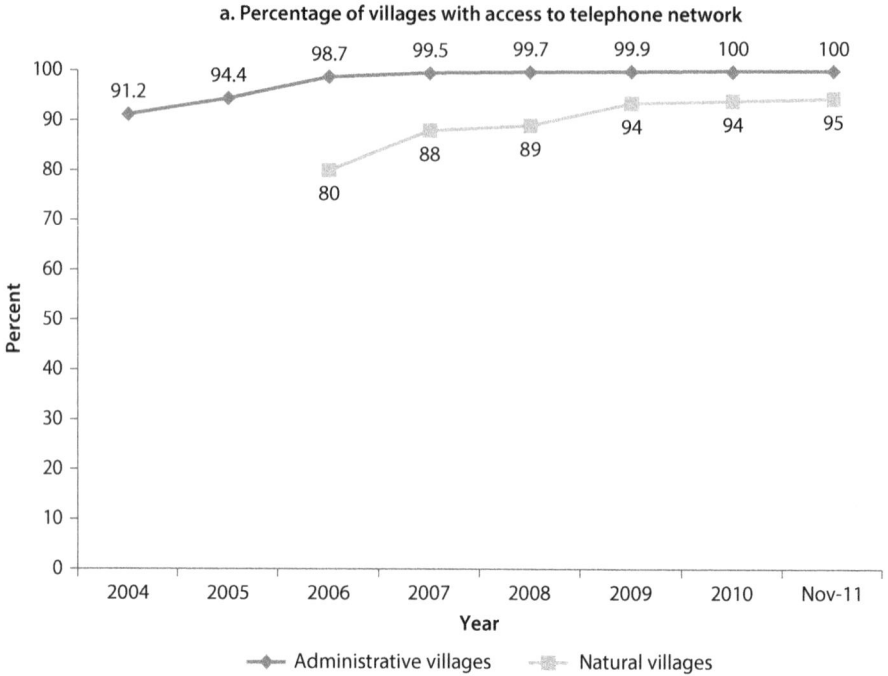

—◆— Administrative villages   —▣— Natural villages

b. Computers per 100 rural households

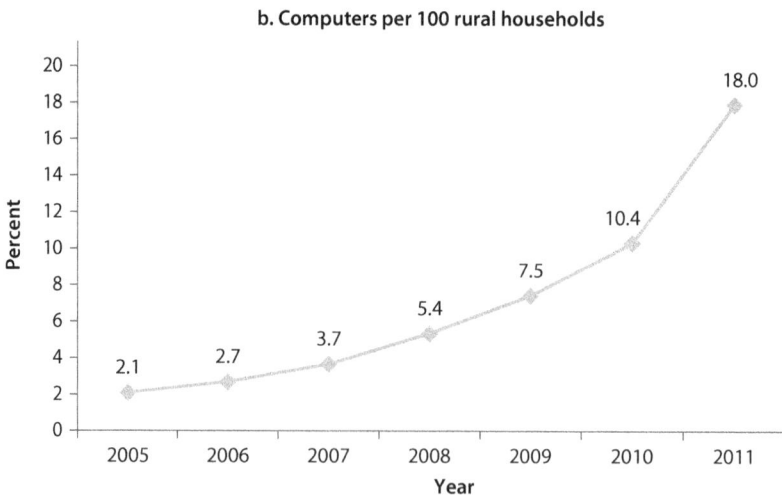

Source: Adapted from National Bureau of Statistics and Ministry of Industry and Information Technology.
Note: Administrative villages have organizing committees and are responsible for a group of natural villages (defined here as villages with more than 20 households).

Box 2.1 Hotline Numbers

China's telecommunication operators are active in rural infrastructure deployment via universal service obligations as well as corporate social responsibility initiatives. High levels of mobile telephone availability in rural areas coupled with low Internet use has led to the deployment of rural information service telephone hotlines using both voice and Short Message Service (SMS) platforms:

- 12582 is an SMS hotline developed by China Mobile. Rural users text both agriculture-related and other questions (for example, health, government benefits, and so on) that are answered by experts in special call centers.
- 12316 is a voice-based service instigated by the Ministry of Agriculture and developed by China Unicom. Similar to 12582, users call in for experts to provide answers to their agriculture-related questions.

In Jilin, the 12316 rural hotline was implemented by the provincial government with the financial and technical support of China Unicom. Some 400 agricultural experts were recruited from universities and research institutes and equipped with mobile phones in return for committing to keeping them on for 24 hours a day. In addition, there is a call center with a dozen staff; if they cannot answer the question, they refer it to one of the 400 experts. Users pay a monthly fee to use the service, and China Unicom reportedly began making a profit with the service after just nine months (Zhou 2007). The 12316 hotline receives an average of 10,000 calls a day. There is also a 12582 SMS platform in the province with more than two million rural users, sending 400 million agricultural text messages on average each year. According to the local government, the two platform services have increased farmer income by more than CNY 4 billion each year.

China Unicom also helped launch the 12316 hotline in Shandong, which has received more than 100,000 calls in the past two years. In Guizhou, voice and text messaging services linked to the province's agricultural information system have 1.3 million users and more than 15 million texts are sent per month.

Rural ICT Interventions in the Three Provinces

Besides the central government, provincial authorities have also carried out a range of activities in support of rural informatization. Examples include development of agricultural information websites, rural information telephone hotlines, television broadcasting including distance education, deployment of public Internet access points, and training. At the suggestion of the Chinese government, rural ICT interventions in three provinces—Guizhou, Jilin, and Shandong (map 2.1 and table 2.2)—were looked at in more detail.

There were several considerations for selecting Guizhou, Jilin, and Shandong as the focus of the analysis. They are at different levels of economic development, with Shandong a relatively developed province, Jilin medium developed, and Guizhou underdeveloped. Geographically, Jilin, Shandong, and Guizhou are located roughly in the east, middle, and west of China, respectively. All three are provinces where agriculture and rural livelihoods are vital to economic and social development. Finally, the three provinces have carried out a number of rural ICT pilot projects.[4]

Information and Communications in the Chinese Countryside
http://dx.doi.org/10.1596/978-1-4648-0204-1

Map 2.1 The Three Provinces Covered by the Program

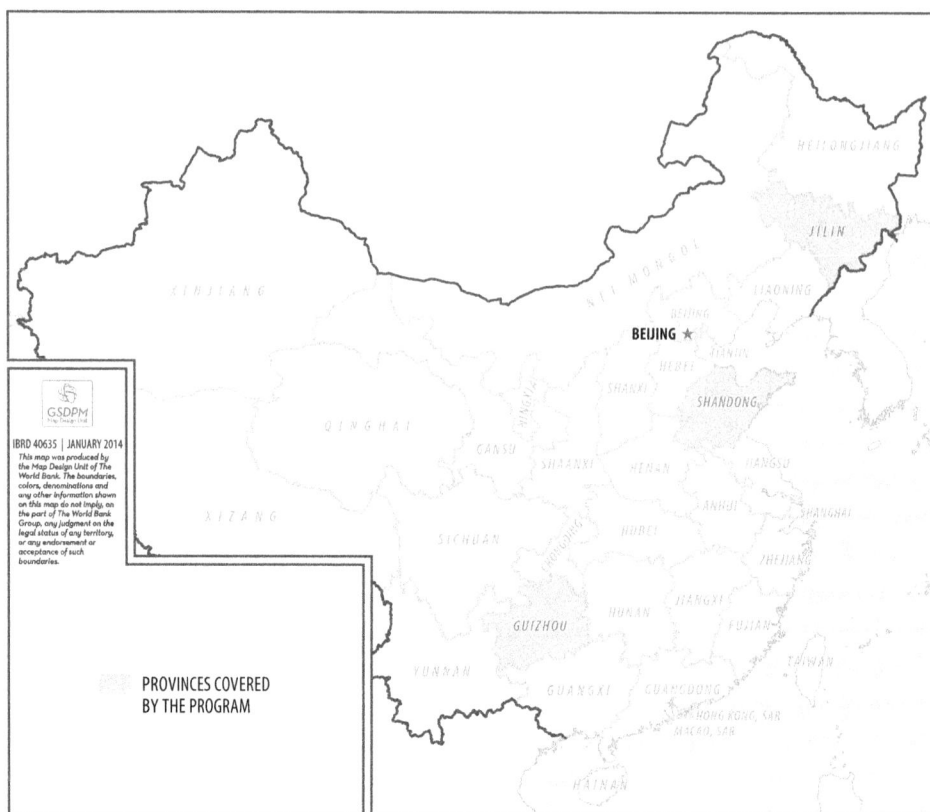

Source: World Bank .

Table 2.2 Basic Indicators in the Three Provinces

Item	Guizhou	Jilin	Shandong
Land area (100,000 km^2)	1.7	1.87	1.53
Population (millions, 2011)	34.7	27.5	96.4
Rural population (%, 2011)	65	47	49
Administrative villages (2006)	19,787	9,441	84,125
Natural villages (2006)	136,994	39,704	107,000
Per capita net Income of rural households (CNY, 2010)	3,472	6,237	6,990
Literacy (age 6+, 2011)	88%	96%	92%
Number of students per 100,000 population (2011)			
Primary	11,749	5,239	6,718
Secondary	3,178	2,874	3,257
Higher education	1,254	2,807	2,191
Administrative villages with access to broadband (%, 2011)	53.8	91.4	100
Internet users per 100 people, 2011 (provincial rank of 31 provinces)	24 (31)	35 (18)	38 (13)

Source: Adapted from National Bureau of Statistics, and China Internet Network Information Center.

Guizhou

The Guizhou Meteorological Bureau has taken a leading role in spreading informatization throughout the province. It manages the Guizhou Agriculture and Economy Network (GAEN) with a team of 50 staff engaged in providing and updating content. The headquarters is also equipped with training facilities.

One key project is the "multi-functional information service station" (Xinhua 2011). These open space centers consist of a reading room, video viewing room, and computers with Internet connection (figure 2.3). They currently exist in 77 villages (see chapter 6). In future the telecenters will be collocated in agricultural parks. The Meteorological Bureau is also developing several other services. This includes smartphone applications on market pricing information and an e-commerce platform for agribusiness being developed in collaboration with Taobao, the Chinese online shopping platform.

Figure 2.3 Multifunctional Information Service Station: Guizhou Province

Computer area

Library section

Video viewing

Source: World Bank

Jilin

The Jilin Agricultural Economics Information Center Company, the business entity designated by the provincial government to develop rural ICT services, has created agricultural content for dissemination over several platforms (see figure 2.4). This includes radio, television, newspaper, websites, text messaging, and telephones covering areas such as product pricing, planting techniques, and other areas of interest to farmers (see chapter 6). It has recently developed a mobile application to exploit the growing availability of smartphones in rural areas.

The province has also launched an e-commerce service implemented under a public-private partnership (PPP) model in partnership with local small and medium enterprises (SMEs). The platform is currently used for purchasing agricultural inputs online. SMEs are provided with computers and Internet connections. Local demand for inputs such as seeds, fertilizer, and pesticides is aggregated in order to obtain lower prices through economies of scale. In addition, free Internet access is made available for farmers at the SME's business location. E-commerce chain shops have been established in some 700 administrative villages with over CNY 50 million ($7.9 million) of trading volume generated through mid-2013.

Figure 2.4 ICT Applications in Jilin Agricultural Economics Information Center Company

Video editing

12582 SMS response center

Website information development

Smart phone application

Source: World Bank.
Note: SMS = Short Message Service.

Shandong

In Shandong province, community centers deploy a number of ICT-based services (see figure 2.5). This includes a "one-stop" counter offering local public services delivery. There are 69 types of permits (for example, planning, real estate) that can be requested, as well as over 500 services. The community center library makes use of smart cards that villagers use to borrow books. There is also an "e-reading room" with 20 personal computers (PCs) where the Internet can be accessed for free. A closed-circuit TV (CCTV) office is used to monitor images from cameras spread throughout the village.

The De Li Si Group, a township food processing business, is involved in the complete cattle value chain including rearing, slaughter, processing, and distribution. It uses radio-frequency identification (RFID) to track the process from farm to supermarket. When the cattle are born, an RFID chip is inserted into their ear. All subsequent events are recorded: slaughter, freezing, transportation, and butchering. Supermarket clients then transfer the RFID information to a bar code that can be scanned by shoppers to check information about the meat.

A fur cooperative in the province provides members with access to the Internet in a computer room at the association's headquarters. The cooperative covers 1,237 households. Members use the Internet to find information about fox breeding as well as for distance education and videoconferencing. It was China's first cooperative to hold a national workshop over the web. The cooperative is developing an e-commerce platform to participate more effectively in the international fur market.

Figure 2.5 ICT in a Community Center, Shandong

One-stop counter

Computer room

Computerized library card

CCTV monitoring room

Source: World Bank.
Note: CCTV = closed-circuit TV.

Information and Communications in the Chinese Countryside
http://dx.doi.org/10.1596/978-1-4648-0204-1

Observations

The Chinese government has introduced a number of interventions to improve rural ICT access. Initiatives come from different central government bodies as well as local administrations. Some are general, targeting an overall expansion goal, such as making broadcast or telephone service available in all villages. Others are more specific and aimed at certain sectors (for example, schools or libraries) or citizens (for example, party cadres, students, or farmers). While the project focused on only three provinces, reports about rural informatization in other provinces paint a similar portrait in terms of interventions and challenges (see box 2.2).

Box 2.2 Rural ICT in Other Provinces

A study of rural areas in **Guangdong** province identified two key ICT projects (Ting and Yi 2012). The Mountainous Region Informatization (MRI) project administered by the Economic and Information Commission (formerly the Department of Information Industry) was a five-year undertaking launched in 2003. It had the goal of providing Internet access to all townships and information services to all villages in the 51 mountainous counties of the province. This included providing funding for establishing "information stations" for farmers to access MRI services and the launching of an SMS agricultural information service by China Mobile. It also involved the creation of agricultural information databases. MRI provided intermediation services through information specialists to assist farmers using the systems as well as China Mobile's hotline. The sustainability of MRI, however, has been an issue. It was hoped that it would become self-funding but the operation has been uneven, with development ongoing where there is strong local support and government financing and diminishing where this is not the case. The Information Express (IE) project was also launched in 2003 and administered by the Department of Science and Technology. The financing and administration differs from the MRI project in that IE has a regular annual budget and is partly carried out by a private contractor. On the other hand, there are similarities in that IE also develops agricultural information platforms and uses a mobile hotline number. The case of Guangdong illustrates the familiar issue of duplicate interventions across many provincial rural ICT programs. This is attributed to China's *tiao-kuai* (line and block) system, where *tiao* refers to the vertical line ministry command chain from central to local governments, whereas *kuai* refers to the different agencies within the same local government. This poses challenges for local officials who are caught between two minds: allegiance to a line ministry or to local governments. This structural problem is reflected in the limited coordination between the two rural ICT projects in Guangdong.

A study of **Sichuan** found certain similarities with rural ICT initiatives in Guizhou province (Liu 2012). Like Guizhou, the Meteorological Bureau plays a leading role in rural informatization. The Rural Economic Information Network (REIN) is an information platform and access network consisting of the provincial headquarters, 21 municipal centers, 188 county centers

box continues next page

Box 2.2 Rural ICT in Other Provinces *(continued)*

and 3,380 service stations in villages and towns. Relevant rural information is collected and disseminated over REIN. China Telecom has also been active in the province through its Information Village project launched in 2008 with the Sichuan provincial government. It aimed to construct 10,000 basic, 1,000 standard, and 100 advanced information villages across the province. China Telecom also began developing agricultural information platforms (Info Countryside) in 2009 to be disseminated over various channels including television, websites, and telephones.

A World Bank team visited **Jiangxi** where it learned that rural informatization in the province is guided by an intergovernmental steering group consisting of the local bureaus of ministries involved in rural informatization. Several towns and villages were visited to witness interventions. This included an information center featuring a library reading room and two PCs with another room dedicated to video viewing. The center is managed by a disabled man, a policy of the province to generate employment for persons with disabilities. In another village, the administrative headquarters includes a kiosk that villagers can use to find out about e-government programs, a private chat room, a play area for users to videoconference with their migrant family members, and a classroom offering distance education. Both formal and informal information workers provide infomediation services in the province. For example, recent college graduates provide assistance on using computers and accessing the Internet; the daughter of a farming family was witnessed browsing the Internet for her parents to obtain agricultural information.

In the absence of a sector regulator and formal universal service fund, the Chinese government has been quite successful in encouraging telecommunication operators to expand infrastructure in rural areas and, more recently, to develop information service platforms. Ministries and provincial governments have administered the development of ICT access centers and information services typically aimed at their target clientele. However, this model has several drawbacks. It has resulted in a degree of duplication with some initiatives from different ministries overlapping in terms of similar types of equipment and services deployed (for example, computers, Internet access and other ICT devices). Emphasis has also been placed on initial project deployment with less attention devoted to follow-up, monitoring, evaluation, and sustainability.

Notes

1. For example the proportion of rural elderly living with their adult children dropped from 70 percent in 1991 to 40 percent in 2006 (Fang, Giles, O'Keefe, and Wang 2012).

2. "…constructing a new socialist countryside is an important historic task in the process of China's modernization…." See: "Consecutive No. 1 central documents target rural issues." *GOV.cn.* 1 February 2010. http://english.gov.cn/2010-02/01/content_1525464.htm

3. The last published statistics on 2G mobile population coverage date from 2008, when China Mobile reported 98 percent coverage. Government officials suggest that current coverage is close to 100 percent (China Mobile Limited 2009, "Annual Report 2008"). The precise number is relevant, given that in a large country such as China, 0.1 percent of the population amounts to over one million people.

4. The State Information Center (SIC) has prepared a background report about rural ICT interventions in the three provinces (SIC 2011).

CHAPTER 3

The Demand for ICT in Rural China

Most official reports about rural informatization in China are focused on monitoring the achievement of targets. There is little information about project follow-up at the grassroots, usage by the assumed beneficiaries, and other outcomes resulting from the interventions. In order to better understand the availability, awareness, and attitudes of and about rural informatization in China, a demand survey was undertaken. The target audience was residents in rural areas of the three provinces (that is, Guizhou, Jilin, and Shandong). The survey was carried out between September and November 2011, covering a statistically representative sample of over 3,000 households in 238 villages. In addition to surveying households, information was collected about infrastructure availability in villages, information and communications technologies (ICTs) in schools, and usage of public Internet centers (see appendix B for additional information about the survey).

Key Findings

The demand survey provided an overview of ICT availability, access, and attitudes in rural areas of three Chinese provinces from various perspectives. Salient observations include the following:

Basic village infrastructure levels are high and television and mobile phones are prevalent in rural Chinese homes. According to the infrastructure audit, all of the villages in the three provinces covered in the survey had electricity and received a mobile telephone signal (see figure 3.1, a). Although over-the-air broadcast channels are not fully available in all villages, this is most likely overcome through satellite dishes. In terms of household possessions, 97 percent had a television and 36 percent had a satellite dish (see figure 3.1, b). Regarding mobile phones, 95 percent of the interviewed households reported having one. Around a fifth of households had a personal computer (21 percent) and 17 percent had a fixed Internet connection.

Low skills inhibit computer and Internet usage. Almost three-quarters of respondents (72 percent) stated they did not use a computer. By far the main reason given was that they did not know how to use it (72 percent). Similarly, 75 percent of respondents did not use the Internet with almost half stating that

Figure 3.1 Village Infrastructure and Household ICT Devices, 2011

a. C. Infrastructure audit.
Does this village have/receive?

b. E.14 Household possessions.
Does your household have?

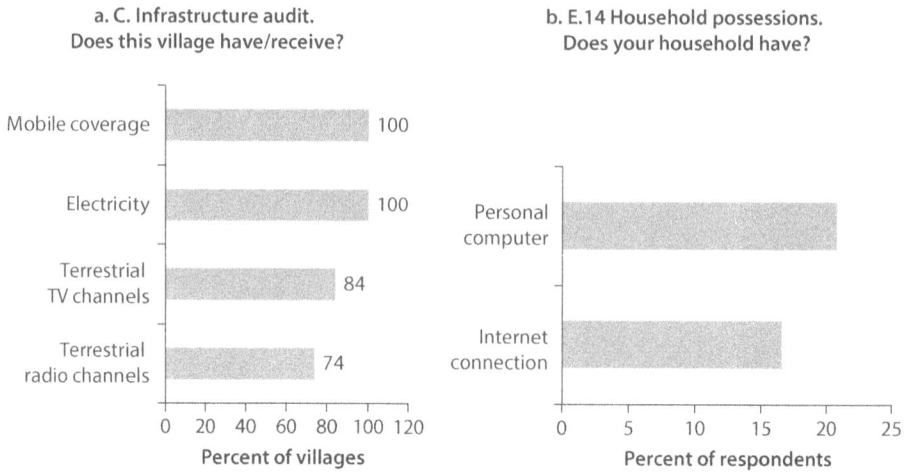

Source: World Bank.
Note: DVD = digital versatile disc; VCD = video compact disc; VCR = video cassette recording.

Figure 3.2 Use and Nonuse of the Internet, 2011

a. E.59 Have you used
the Internet?

b. E.60 If you do not use the Internet,
what is the main reason?

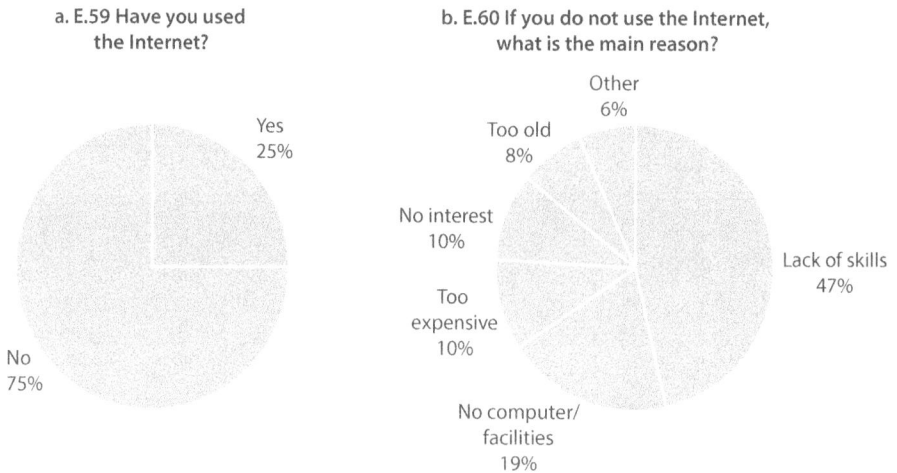

Source: World Bank.
Note: Asked of household member present at time of survey. N=3,060.

the reason was lack of skills (see figure 3.2). There are few opportunities for rural households to receive training. Of schools with Internet access, only a quarter (26 percent) reported making their facilities available to the local community, and only 9 percent of public ICT facilities reported that they offered training.

High ownership of mobile phones. The survey found individual ownership of mobile phones to be 85 percent, with over half of individuals without their own mobile reporting that they did not have one because they could use someone else's or they had no need (see figure 3.3, a). About half of mobile phone owners

Figure 3.3 Individual Mobile Phone Ownership and Use, 2011

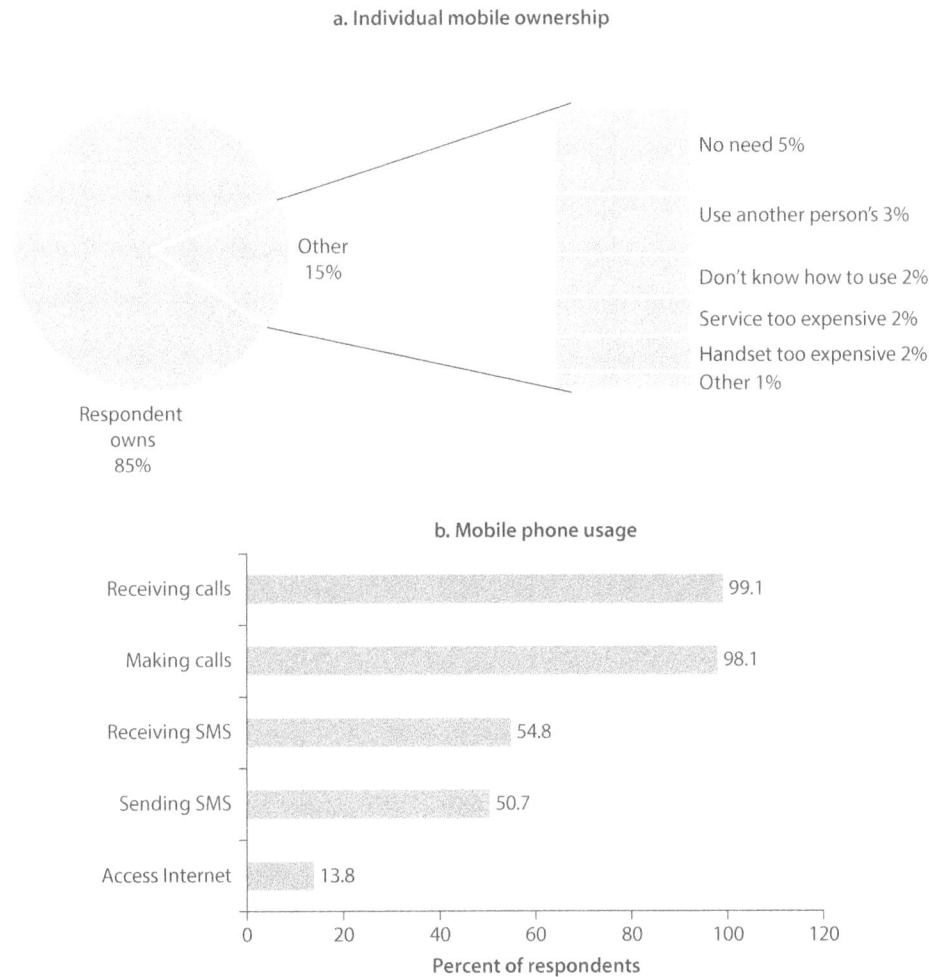

a. Individual mobile ownership

No need 5%

Use another person's 3%

Other
15%

Don't know how to use 2%

Service too expensive 2%

Handset too expensive 2%
Other 1%

Respondent
owns
85%

b. Mobile phone usage

Receiving calls	99.1
Making calls	98.1
Receiving SMS	54.8
Sending SMS	50.7
Access Internet	13.8

Percent of respondents

Source: World Bank.
Note: SMS = Short Message Service.

reported sending text messages, and some 13 percent used the Internet from their mobile phone (see figure 3.3, b). Average monthly mobile phone service expenditure was 13 percent of income, with users willing to devote up to 18 percent of their income to mobile services.

Affordability is not the main barrier to rural ICT access but remains a concern. The main reasons households do not have Internet is because of a lack of skills or because they had no computer (accounting for 66 percent). The cost of access ranked third, with 10 percent stating that this was the reason they did not have the Internet. It is worth noting that while 19 percent of those without the Internet stated that it was because they did not have a computer or Internet availability, 24 percent of those without a computer replied that they did not have one because it was too expensive.

Information and Communications in the Chinese Countryside
http://dx.doi.org/10.1596/978-1-4648-0204-1

Figure 3.4 Main Uses of the Internet, 2011

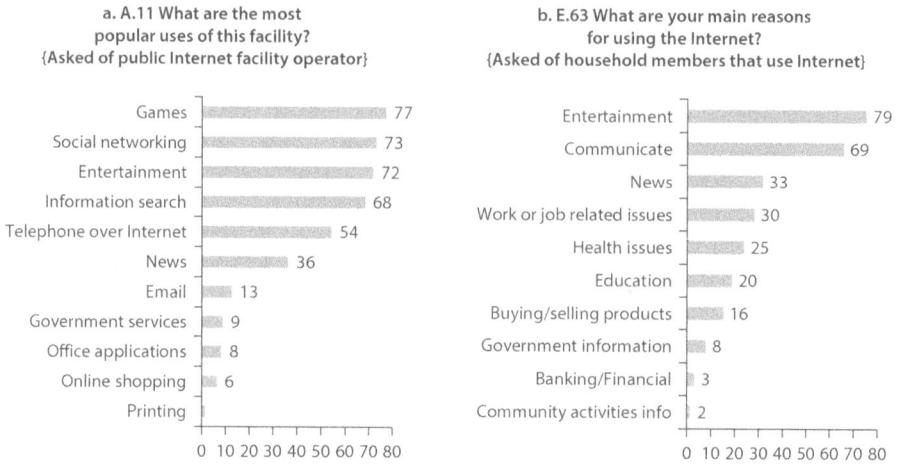

a. A.11 What are the most popular uses of this facility? {Asked of public Internet facility operator}	b. E.63 What are your main reasons for using the Internet? {Asked of household members that use Internet}

Games — 77
Social networking — 73
Entertainment — 72
Information search — 68
Telephone over Internet — 54
News — 36
Email — 13
Government services — 9
Office applications — 8
Online shopping — 6
Printing —

0 10 20 30 40 50 60 70 80

Entertainment — 79
Communicate — 69
News — 33
Work or job related issues — 30
Health issues — 25
Education — 20
Buying/selling products — 16
Government information — 8
Banking/Financial — 3
Community activities info — 2

0 10 20 30 40 50 60 70 80

Source: World Bank.
Note: In the left chart, "Entertainment" refers to watching videos and downloading songs, and "Office applications" refers to word processing, spreadsheet, and so on. In the right chart, "Entertainment" includes games, and "Communicate" refers to relatives and friends (for example, chat, social networking, and so on).

The primary use of the Internet in rural areas is to meet villagers' demand for entertainment. Interviews with the managers of public access points indicated that the top three activities performed by users were playing games (77 percent), social networking (73 percent), and entertainment (for example, watching videos or downloading songs) (72 percent) (see figure 3.4, a). Sending e-mail, using business applications, shopping online, and printing were seldom performed. Household Internet users indicated their main online usage was entertainment and communication (figure 3.4, b). This suggests that the primary use of the Internet in rural areas is to meet villagers demand for amusement and social networking.[1]

Shared Internet access is mainly via private Internet cafés. After home, household respondents said that an Internet café was the next place where they mainly used the Internet (see figure 3.5, b). According to the public Internet facility survey, only a quarter of the surveyed sites were in the village surveyed (see figure 3.6, a) (if there was no public Internet facility in the village, then interviews were carried out in the nearest facility). More than four out five of the public Internet facilities surveyed were privately operated (see figure 3.6, b).

Internet cafés are mainly used by young men. Private Internet cafés tend to be in towns and to be quite large, with an average of 40 computers per facility. According to the staff of Internet cafés, users are mainly men and the young (see figure 3.7). Further, the most popular use was gaming and entertainment, reported by over 70 percent of respondents. In contrast, business use of office applications, printing, and scanning was reported to be less than 10 percent.

Despite the potential of public access points to provide training, only 15 percent of household respondents reported that they had received ICT courses there. Some 25 percent of households reported using the intermediation of an "information worker" at a public access facility.

Information and Communications in the Chinese Countryside
http://dx.doi.org/10.1596/978-1-4648-0204-1

Figure 3.5 Public Internet Facilities, Availability and Usage, 2011

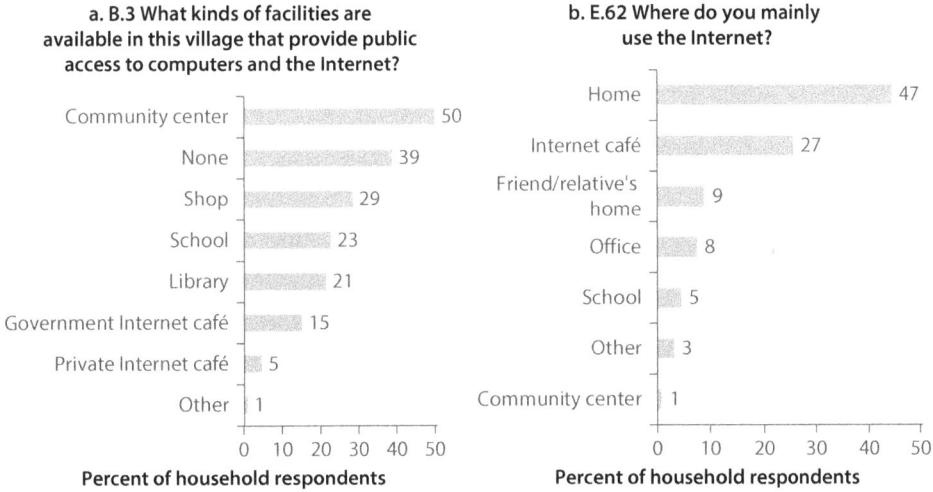

a. B.3 What kinds of facilities are available in this village that provide public access to computers and the Internet?

Facility	Percent
Community center	50
None	39
Shop	29
School	23
Library	21
Government Internet café	15
Private Internet café	5
Other	1

Percent of household respondents

b. E.62 Where do you mainly use the Internet?

Location	Percent
Home	47
Internet café	27
Friend/relative's home	9
Office	8
School	5
Other	3
Community center	1

Percent of household respondents

Source: World Bank.

Figure 3.6 Public Internet Facilities, Location and Management, 2011

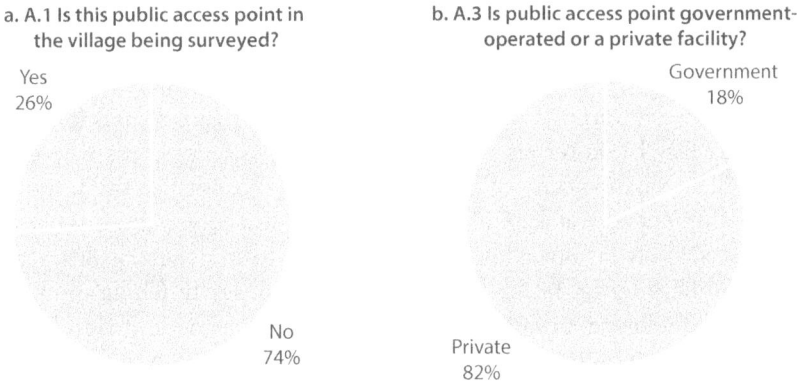

a. A.1 Is this public access point in the village being surveyed?

Yes 26%
No 74%

b. A.3 Is public access point government-operated or a private facility?

Government 18%
Private 82%

Source: World Bank.

Figure 3.7 Private Internet Café, Jilin Province and Distribution of Users, Public Access Facilities, 2011

A6. What percentage of users come from these groups?

Group	Percent
Male	82
Youth	73
Migrant workers	19
Farmers	19
Female	17
Business people	10
Government workers	8
Students	6
Elderly	5
Housewives	2

Source: World Bank.

Information and Communications in the Chinese Countryside
http://dx.doi.org/10.1596/978-1-4648-0204-1

Figure 3.8 Sources of Information and Use of Internet Services, 2011

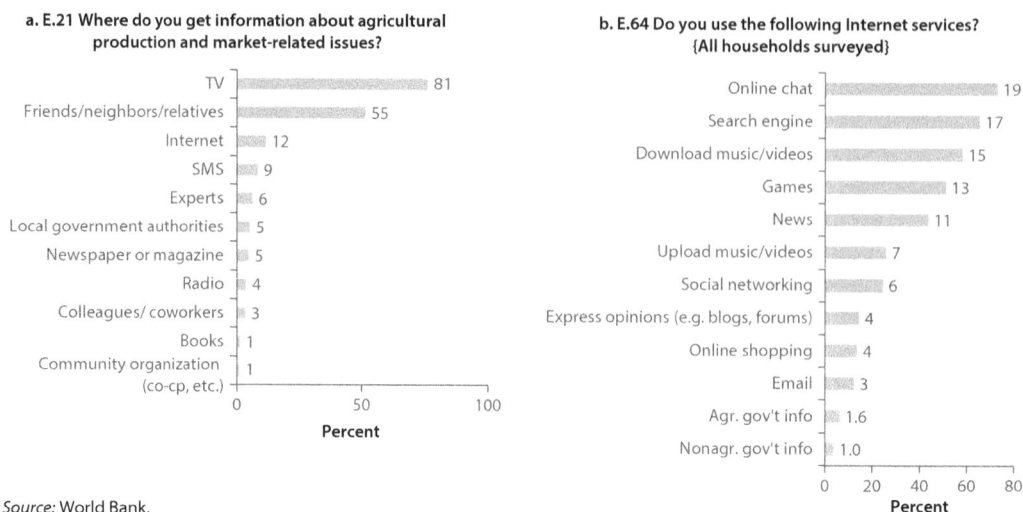

a. E.21 Where do you get information about agricultural production and market-related issues?

- TV — 81
- Friends/neighbors/relatives — 55
- Internet — 12
- SMS — 9
- Experts — 6
- Local government authorities — 5
- Newspaper or magazine — 5
- Radio — 4
- Colleagues/ coworkers — 3
- Books — 1
- Community organization (co-cp, etc.) — 1

(x-axis: 0, 50, 100; Percent)

b. E.64 Do you use the following Internet services? {All households surveyed}

- Online chat — 19
- Search engine — 17
- Download music/videos — 15
- Games — 13
- News — 11
- Upload music/videos — 7
- Social networking — 6
- Express opinions (e.g. blogs, forums) — 4
- Online shopping — 4
- Email — 3
- Agr. gov't info — 1.6
- Nonagr. gov't info — 1.0

(x-axis: 0, 20, 40, 60, 80; Percent)

Source: World Bank.
Note: SMS = Short Message Service.

Limited "information circle". The vast majority of household respondents obtained information from friends and relatives or television (see figure 3.8, a). Provincial authorities have devoted considerable resources to develop agricultural websites, yet less than 2 percent of all rural household respondents reported using the Internet to access online government agricultural information (see figure 3.8, b). Entertainment was the primary use for those who used the Internet. The vast majority of those interviewed did not use the Internet. The main reason stated was lack of skills. The ways to acquire those skills are limited: Most schools do not allow the community to use their facilities and hardly any public Internet facilities provide training. Despite China's impressive strides in deploying infrastructure outside urban areas and developing increasing amounts of online information aimed at farmers, many rural inhabitants are not making effective use of it and are missing out on the potential transformative benefits. As one rural expert noted: "For these services to become effective, ways need to be devised to lead farmers to assert usership over these resources and to sustain their interest in using them. This is far more challenging than delivering the resources into the boundaries of farmers' information worlds" (Yu 2010).

Observations

Owing to the lack of detailed information about the demand side of ICT interventions in rural areas of China, the project carried out a survey to get a better understanding of rural ICT availability and usage in the three provinces.

The survey found a generally high level of basic infrastructure in villages with all of those studied having electricity and mobile coverage. Households also had a high level of television and mobile phone ownership. Levels of computer penetration and Internet use, however, were generally low. Further, there was little evidence that mobile phones are currently substituting for Internet access to a significant degree. The main barrier to greater computer and Internet take-up is a lack of skills or a perceived lack of relevance.

The availability of computer training is limited. In general, venues that might normally offer training courses either do not exist in villages, lack the necessary equipment, or have restrictive access to the facility. Internet cafés are available in townships but tend to be geared toward entertainment and dominated by young men. Most do not provide training but even if they did, they do not offer a suitable environment for some groups, such as women or older people. Further, the migration of working-age people to urban areas means that there is less opportunity to acquire digital training from family members.

Affordability is generally not a barrier to rural ICT access. Overall, about a quarter of respondents without a computer said that it was too expensive. Regional differences are significant with less than 10 percent of respondents in Shandong province stating that the cost of computers was a barrier compared to almost 40 percent in Guizhou, reflecting different levels of income in the two provinces.

Digital training is sorely needed in rural areas in order to increase access to ICTs. It is not realistic to assume that ICT access will improve by the simple provision of infrastructure. There also needs to be a corresponding formal training platform that is available to all in the community. It should be structured as part of a program with the components tailored to the abilities and demeanor of different groups (for example, self-employed, women, older people, and so on) and widely publicized. Targets should be established for the number of villagers trained including some type of certification system. The cost of ICT devices and access is an issue for some rural residents and could be ameliorated through initiatives similar to the home subsidy program mentioned earlier or making convenient shared access available through inclusive community venues.

Note

1. This is consistent with nationwide studies: "Internet entertainment functions have great attraction to rural internet users; online music, online video, online games and chat are primary purpose in using the internet of rural users." Some 83 percent of rural inhabitants listed to online music in 2009, 70 percent played online games and 57 percent watched online games (Yu and Qin 2011).

CHAPTER 4

ICT in Rural Schools

Primary and secondary schools can make a valuable contribution to China's rural informatization by providing a venue where students can learn how to use computers and the Internet. Schools could also provide after-hours access to information and communications technology (ICT) facilities for the local community. This is particularly important, given that almost half of rural residents who do not use the Internet cited a lack of skills as the reason why. This chapter reviews ICT initiatives for schools, particularly in rural areas, and also draws on school data from the three provinces carried out as part of the ICT demand survey. The focus here is on the potential of schools to increase accessibility and impart basic computer skills rather than on the use ICTs for learning and other impacts.[1]

National Initiatives for ICTs in Schools

Primary (grades 1–6) and junior secondary (grades 7–10) schools are the most prevalent in China's rural areas and attendance is compulsory. Senior secondary schools, vocational institutions, colleges, and universities tend to be clustered in larger towns and cities. The primary and junior secondary education system consists of over 281,000 schools, 145 million students, and 8.6 million teachers (see table 4.1). The majority of primary schools (68 percent) are located in rural areas. However, the share of primary school students in rural areas is smaller (37 million or 38 percent of the total) owing to smaller class sizes. The proportion of schools, students, and teachers is smaller in rural areas for junior secondary schools because attrition rates increase, and also because there are fewer higher-level schools in rural areas. Only 5 percent of senior secondary schools are located in rural areas.

Since early 2000, several initiatives led by the Ministry of Education have promoted ICT access in schools in China. This includes three launched in 2000 (Chen 2003):

- The School Connection Project supported Internet connectivity and ICT application in primary and secondary schools. One target included connecting 90 percent of primary schools to the Internet by 2010.

Table 4.1 Number of Primary and Junior Secondary Schools, Students, and Teachers, 2012

		Total	Urban	Counties & Towns	Rural
Primary	Schools	228,585	26,146	47,431	155,008
		100%	11%	21%	68%
	Students	96,958,985	26,884,287	33,549,812	36,524,886
		100%	28%	35%	38%
	Teachers	5,121,626	1,254,960	1,703,810	2,162,856
		100%	25%	33%	42%
Junior secondary	Schools	53,216	10,932	22,876	19,408
		100%	21%	43%	36%
	Students	47,630,607	14,410,251	23,479,363	9,740,993
		100%	30%	49%	20%
	Teachers	3,504,363	1,021,532	1,701,220.00	781,611.00
		100%	29%	49%	22%
Total	Schools	281,801	37,078	70,307	174,416
	Students	144,589,592	41,294,538	57,029,175	46,265,879
	Teachers	8,625,989	2,276,492	3,405,030	2,944,467

Source: Adapted from Ministry of Education, "Educational Statistics in 2012".
Note: Teachers refer to full time.

- Training Guidance for Teacher Training about Information School, published by the Teacher Education Department of the Ministry of Education, urged all primary and secondary school teachers to participate in professional development to learn how to use information technology.
- Popularizing ICT Education in primary and secondary schools launched in 2000 whereby schools were requested to launch courses for students learning how to use computers.

The Distance Education Project for Rural Schools project was launched in 2003 to reduce imbalances between urban and rural schools, particularly those located in the central and western parts of the country. Three models for accessing information were implemented depending on the type of school. Digital versatile discs (DVDs), DVD players, and television sets were to be provided to 110,000 primary schools in sparsely populated villages; DVDs, DVD players, television sets, and satellite dishes were to be provided to 384,000 primary schools in townships; and computer labs were to be established in 37,500 junior secondary schools. By 2007, the project had been implemented in 80 percent of the schools (McQuaide 2009).

These initiatives have led to increasing informatization of schools, progressing from providing basic equipment and access, to connecting classrooms, and finally to enhancing connections to students. According to the Ministry of Education, every "complete" school (schools that have the full complement of grades for its level) has Internet access.[2] However there remain noticeable

Table 4.2 Number of Computers per 100 Students and Percentage of Digital Campuses, 2010

	Computers per 100 students			Digital campus (%)		
	Total	Urban	Rural	Total	Urban	Rural
Primary	4.1	7.2	3.5	15.9	64.7	12.6
Junior secondary	6.4	7.8	6.0	46.4	71.5	42.6

Source: Zeng, Huang, Zhao, and Zhang 2012.
Note: Digital campus refers to schools that have local area network on campus and Internet access.

differences in levels of computerization and networking between primary and junior schools and urban and rural regions (see table 4.2). By 2010, there were 4.1 computers per 100 students in primary schools and 6.4 in junior secondary; 16 percent of primary schools had a local area network compared to 46 percent of junior secondary schools. While there were 7.2 computers per 100 students in urban primary schools the corresponding figure for rural ones is less than half that rate.

The *National Plan for Educational Informatization (2011–20)* was published in 2012. Among other things, it calls for complete broadband coverage of all schools and the establishment of a platform for all schools to access online educational content.[3]

Findings from Surveys

The demand survey included a questionnaire targeted at primary and secondary school principals in the villages. Results were obtained from a total of 110 rural schools in the three provinces. Some 45 percent of villages had a school, almost all primary schools. All of the surveyed schools had electricity (figure 4.1). In terms of computers, 76 percent of the schools had personal computers (PCs). The main reason for not having PCs was "no budget" in Jilin and Guizhou, whereas all schools had computers in Shandong.

The survey found that only 63 percent of schools could afford an Internet connection. Some schools did not realize the "need" to have Internet access at school. Almost 60 percent of schools with Internet access did not allow students to use Internet in the school. Moreover, 73 percent of schools with Internet access did not allow villagers to use it.

In terms of ICT usage in teaching, the average was 27 percent, although there were differences among the three provinces. Shandong, with the highest PC possession and Internet connection rate in schools, also had the highest rate of ICT usage in teaching (42 percent). In terms of ICT education for teachers, the survey found that almost all teachers received training: 81 percent of teachers took formal information technology (IT) literacy training courses offered by the Ministry of Education; another 29 percent took informal training.

Information and Communications in the Chinese Countryside
http://dx.doi.org/10.1596/978-1-4648-0204-1

Figure 4.1 Equipment and Facilities in School Surveyed by Province, 2011

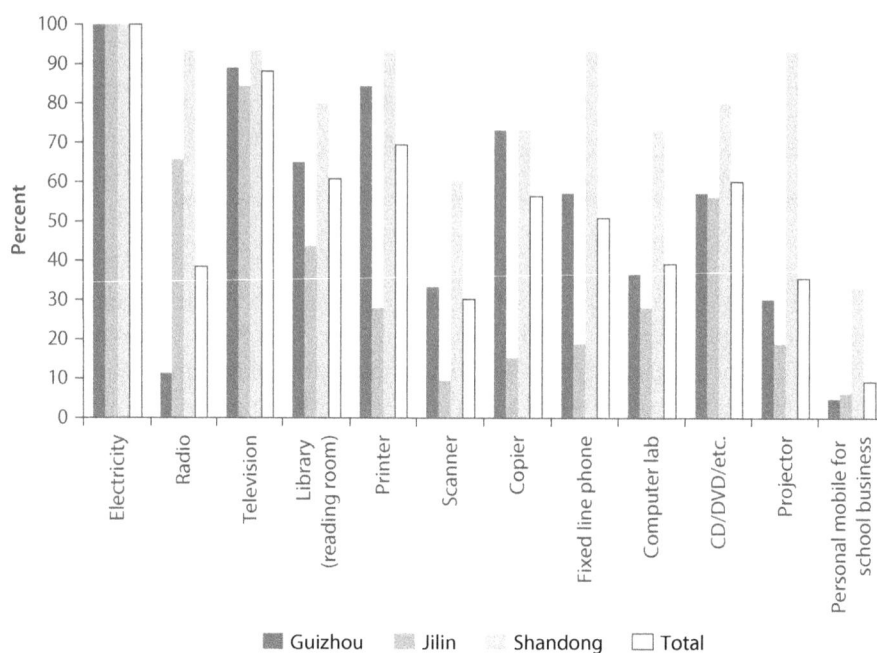

Source: World Bank.
Note: CD = compact disc; DVD = digital versatile disc.

Table 4.3 Computer Use in Beijing and Shaanxi Public Schools, 2010

	Urban students (%)	Rural students (%)
In the school rooms of the students		
Use computer at school	88	69
For students who use computers at school, they have at least one computer class every week	100	72
40 minutes or more per computer class	100	78
What is learned in ICT class		
Learned how to turn on /off the computer	100	84
Learned how to use keyboard	100	76
Learned how to use the mouse	100	80
Learned how to type Chinese	100	68
Learned how to draw	100	70
Used educational software	90	36
Learned about computer hardware	90	39
In the homes of the students		
Have some type of computer (laptop, desktop, and so on)	80	10
Can access Internet at home	73	5

Source: Zhang, Lai, Shi, Boswell, and Rozelle 2013.
Note: ICT = information and communications technology.

It is interesting to contrast the demand survey findings with another survey carried out by the Rural Education Action Program (REAP) comparing differences in computer availability and use between urban and rural primary schools in Beijing and Shaanxi province in 2010 (see table 4.3) (Zhang, Lai, Shi, Boswell, and Rozelle 2013). While 88 percent of urban students used a computer at school, the corresponding figure in rural schools was 69 percent. Some 80 percent of urban students had a computer at home and 73 percent had Internet access; the corresponding figures for rural students were 10 percent and 5 percent.

Observations

China's high enrolment rate—over 100 percent at primary and 81 percent at secondary level[4]—suggests that if ICTs are available in schools, they would be accessible to most of the school-age population. However, there is a gap between ICT availability and use in urban and rural schools. This gap tends to narrow with the level of schooling, partly a reflection that from junior secondary upwards, schools are more likely to be located in towns and urban areas where there are greater resources for ICT. Higher levels of access among secondary school students are reflected in national Internet usage surveys, where 84 percent of those between the ages of 10 and 19 were estimated to be using the Internet in 2012 compared to only 13 percent for those between 6 and 9 years old (figure 4.2, left). It is not clear how much the high level of access for the 10–19 age group is because of use in schools or because of additional education from junior secondary schools (figure 4.2, right). It suggests that if the trend continues, future generations of Chinese will be at least basically computer literate, and the focus for training and skills development should shift to older citizens.

Figure 4.2 Percentage of Age Group Using the Internet and Distribution of Internet Users by Educational Attainment, 2012

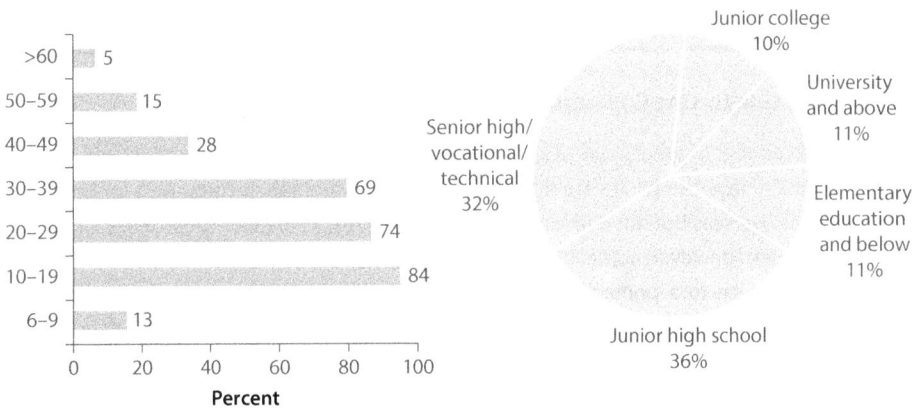

Source: Adapted from CNNIC and China Bureau of Statistics.
Note: The left chart is derived by taking the reported share of Internet users by age in 2012 and dividing by the 2011 population of that age group. CNNIC = China Internet Network Information Center.

Leveraging school ICT facilities for student and community use could help to improve digital literacy in China. Even though computers and the Internet may be available in schools, students and villagers are often not allowed to use them. The demand survey found that 58 percent of schools did not allow students to use PCs and 81 percent did not allow students to use the Internet, presumably reserved for administrative purposes. Only 26 percent of schools with the Internet allowed the local community to use it. Thus, the potential of schools as a venue for spreading digital literacy is currently constrained. Legal and other barriers restricting community use of school computer labs need to be overcome.

The number of computers in schools is relatively limited. Even in urban junior secondary schools, on average there is only one computer for about 13 students. This limits the time available to do research and homework and the ability to use educational software. Further, in rural areas, few homes have computers for students to use. Some countries have launched "one computer for every child" programs involving the distribution of laptops or tablets to children. A potential side benefit of these programs is that, when children bring their computers home, parents and grandparents can also use it. One of the most successful programs has been in Uruguay where all school-age children received a laptop computer. The Uruguayan program aimed to democratize computer use by first distributing the laptops in rural areas (Batista 2009). Surprisingly, although China is the leading manufacturer of laptop and tablet computers for one-to-one projects in other countries, there is no such program in the country on a national scale (see box 4.1).

Resources need to be increased for schools in rural areas to reduce the gap with urban schools. In the ICT demand survey, 77 percent of schools reported that they did not have PCs because of lack of funding. Broadband also needs to be extended to schools and the cost of access needs to be affordable. One third of schools reported that Internet service was not available in their village, 65 percent reported they could not afford the Internet, and one-quarter of those with the Internet were using dial-up connections.

Box 4.1 One-to-One Computing in China

Unlike a number of other countries, there is no formal national one laptop per child program in China. Chinese companies are manufacturing laptop and tablet computers for large-scale programs overseas but not domestically (Montlake 2012). There have been a few small-scale interventions, however. Some 1,000 laptops were distributed by the One Laptop per Child (OLPC) program to a primary school in Sichuan province following the devastating 2008 earthquake.[5] In Beijing, laptops were distributed to 150 grade three students in migrant schools as part of an impact assessment (Mo, Swinnen, Zhang, Yi, Qu, Boswell, and Rozelle 2013). The study found that, six months after the intervention, the program had improved computer skills by 0.33 standard deviations and math scores by 0.17 standard deviations. It concluded that laptop programs could be an effective way of reducing the digital divide.

One positive finding from the school survey is that about 90 percent of schools reported having teachers trained in ICT. If access to computers and the Internet could be extended—by allowing students and villagers to use existing equipment, and making computers and Internet access more affordable for schools—teachers are available that could then offer basic ICT training.

Notes

1. For a list of research carried out on ICTs in Chinese schools, see: Trucano 2012.
2. World Bank meeting with Ministry of Education, October 21, 2013.
3. http://www.unescobkk.org/fileadmin/user_upload/ict/Workshops/amfie2012/presentations/Du_Zhanyuan.pdf
4. Data relate to 2011 and refer to gross enrolment ratio defined as number of students, regardless of age, expressed as a percentage of the population of the official education age. The figure can exceed 100 percent due to the inclusion of overaged and underaged students because of early or late school entrance and grade repetition. Source: http://data.worldbank.org (SE.PRM.ENRR and SE.SEC.ENRR).
5. http://www.olpc.asia/downloads/story-of-olpc_2up_EN.pdf

CHAPTER 5

Public Library Landscape

Libraries are a logical partner for rural informatization because of their community-facing infrastructure. The Chinese government has undertaken a number of rural library informatization projects. The challenges for township and village public libraries include uncertain and insufficient investment leading to unsustainable development and sometimes closures. Further, the availability of digital resources in rural public libraries and digital skills among rural librarians are limited.

An overview study of the Chinese public library system was undertaken in order to take stock of the existing situation and consider potential operating models to provide public information access through local libraries and library-like institutions.[1] In addition, a survey was undertaken in the three provinces to explore library services and usage.

Chinese Public Library System

The Ministry of Culture is responsible for public library policy at all levels. The Ministry of Culture also administers the National Library, but there is no subordinate relationship between the National Library and public libraries. Since 2005, the Chinese government has issued a series of general and specific policies to guide and facilitate inclusive cultural development. In these documents, the government aims to develop a public culture service to ensure equal access and contribute to the wellbeing of all citizens. Taking into account uneven economic and social development, the biggest challenge is promoting the construction of cultural facilities in rural areas where financial and human resources are limited.

In general, Chinese public libraries follow the public administrative system (that is, the Central Committee—provinces—regions [prefecture-level cities]—counties, cities—towns—administrative villages). Local governments at different levels are in charge of construction, operation, maintenance, and management of public librarianship. Expenditure on public libraries is included in the financial budget of the corresponding level of government; however, there is no legal requirement for local governments at different levels to set up a public library within their jurisdictions. The curator of the library is appointed by the cultural division of respective level governments, which are responsible for the operating

http://dx.doi.org/10.1596/978-1-4648-0204-1

expenses of the library. In terms of linkages, public libraries at the next higher administrative level are in charge of providing professional guidance for the public libraries at the next lower level.

Within each region, different levels of local government have the responsibility to maintain public libraries (including annual investment, human resource expenditure, building and equipment, and so on). This traditional model has led to a waste of resources, overlapping investment, and loose relationships among public libraries. Especially for lower-level public libraries, a lack of funding has hampered their development and has sometimes resulted in closures.

A new model has been emerging which aims to remedy this system. It seeks to change the traditional autonomous status of each library. Under this model, the main provincial library becomes the "core" that channels investment from different levels of government.[2] It coordinates the allocation of funding, staffing, and information resources in order to improve the effectiveness of expenditure on libraries. This model has been promoted by government as an example for widespread adoption. In the national culture development plan, promoting this type of main library-branch library relationship is regarded as a major goal across the whole country.

National Projects
A series of projects led by the government has sought to enhance information technology application in public libraries in rural areas of China.

Comprehensive Cultural Station Construction Project
The project, also known as the multifunctional information service system, is led by the Ministry of Culture and aims to extend public library services by assisting township governments in developing comprehensive cultural stations. These stations are the physical hosts of activities, programs, and projects such as book readings, TV and film shows, training, entertainment, physical exercise, support for children, and so on. For example, the Cultural Information Resources Sharing Project and the Farmers' Book House Project (see below) both depend on the comprehensive stations. The project is mainly designed to help establish buildings for the cultural stations, including a multifunctional hall, reading room, training classroom, and a cultural information resources sharing service room. It should be noted that the comprehensive culture stations are only located at the township level.

Cultural Information Resource Sharing Project
The project, initiated in 2002 by the Ministry of Culture, aims to make the best use of modern technology to digitize Chinese cultural information resources accumulated over thousands of years as well as those in modern society relevant to life. Cultural Information Resource Sharing Project (CIRSP) is a platform based on the national communication network using the Internet and satellite broadcasting. Digital information is also provided on digital versatile disc

(DVD) and is broadly used in the project because of its mass distribution capability, wide accessibility, and lack of reliance on network transmission systems (Wu 2012).

The project aims to build and share cultural information resources across the country by establishing information resource centers, network centers, and an information resource network transmission system which covers all provinces, autonomous regions, directly administered municipalities and most prefecture-level cities, counties (and county-level cities) as well as some towns, villages, and residential districts. The structure of the CIRSP is national center (located in the National Library)—provincial center (one at each provincial public library)—prefecture/county subbranch center (at each prefecture-level and country-level public library)—township service station (at township cultural stations)—village-level service station (at village community centers).

By the end of 2013, the project had established almost 900,000 service points, including 33 at the provincial level (100 percent coverage), 333 at the prefecture level (100 percent coverage), 2,843 at the county level (99.7 percent coverage), 2,555 at the township level (89 percent coverage, and 602,000 service stations (99 percent coverage) at the village level (figure 5.1). In addition the project was available to 250,000 schools in rural areas. It is reported that 960 million people were considered to have benefited from the project[3] although, given statistics on library use, this figure undoubtedly refers to theoretical coverage. The CIRSP also sponsors the Public Electronic Reading Room Construction Plan (see below).

Figure 5.1 Scope of CIRSP

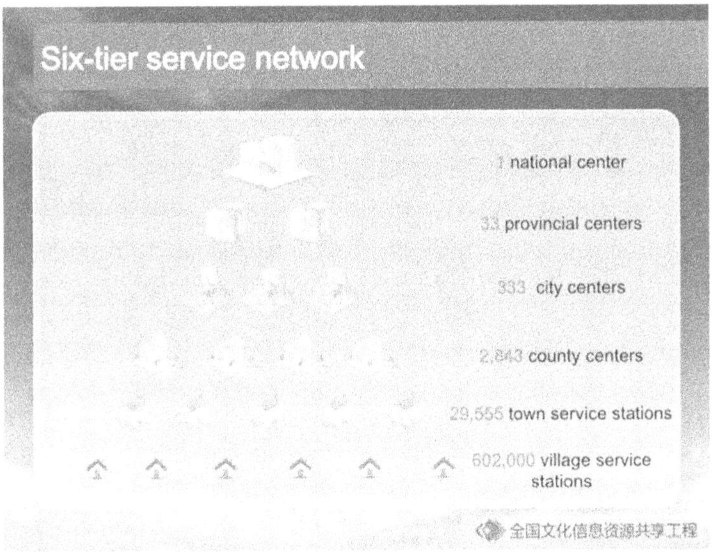

Source: Public Culture Development Center. 2014. "Status of the National Cultural Information Resources Sharing Project and the next Focus" presented at the Workshop on the Sustainable Development Models of China Rural ICT, January 22, 2014, Beijing.

Digitized information from libraries, art galleries, and museums and from the fields of TV, education, science and technology, and agriculture is incorporated in the CIRSP content resources. By the end of 2011, the volume of such digital resources had reached 136 TB, including 70,000 hours of video programs and about four million e-books.

Farmers' Book Houses

These are established to meet the cultural needs of farmers in administrative villages, managed by the farmers themselves, equipped with free books, periodicals, and audiovisual resources on practical subjects. The project began in 2007 and covers the whole of China. A Farmers' Book House has some of the basic functions of a public library, including lending. Each one covers over 50 square meters and is intended to hold 1,500–5,000 volumes, 30 newspaper and periodical titles, 100 DVDs and compact discs (CDs), and is equipped with players and furniture for reading and study.

The central government funds half the cost of each house in mid-China, and 80 percent in west-China at the level of CNY 20,000 ($3,170) for each house. Between 2007 and 2012, central and local government invested CNY 12 billion ($1.9 billion). More than 600,000 Farmers' Book Houses have been set up throughout China, and almost every village has one. More recently, some provinces are seeking to develop *Digital* Farmers' Houses using multimedia, Internet, satellite and cable TV, and web technologies.

Public Electronic Reading Room Construction Plan

Since February 2012, the Public Electronic Reading Room Construction Plan has been underway across the whole country. The project relies on the service network of the CIRSP and the digital resources of the CIRSP and the National Library of China, along with the physical buildings of the Comprehensive Culture Station Project. The aim of the project is to provide a free, content-safe and clean, service-normative Internet service space for the public, free of charge, especially aimed at minors, seniors, and rural migrant workers in cities. This project aims to provide a free electronic reading room in all townships, subdistricts, and communities including the integration of officially approved private Internet bars. Each reading room is required to be equipped with at least 40 square meters of space, 10 computers, a local area network with storage, and Internet bandwidth of not less than 2 Mbps.

Free to All

In January 2011, the official document, *Working Opinions on the Promotion of the National Art Gallery, Public Library, Cultural Centers and Stations Free to All*, issued by the Ministry of Culture and the Ministry of Finance, required use of public spaces including public libraries to have no charge for the public (including services such as reading, book lending, searching, and reference, nonprofit lecture, and so on). The government provides an annual subsidy of CNY 500,000 for each public library at the regional level, CNY 200,000 for each county library, and CNY 50,000 for every Comprehensive Cultural Station. The subsidy from central government varies between different parts of China.

For the mid-China area, each level of library and station receives a 50 percent subsidy separately from central and local government; for the less developed regions in the west, the subsidy from the central government rises to 80 percent, and local government would provide the remainder. For the eastern areas, local government offers the subsidy. In 2011, the subsidy from central government reached CNY 1.8 billion ($285 million).

Modern Distance Learning of National Party Cadres in Rural Areas Project
This is another national project initiated by the Central Committee of the Communist Party that offers one computer with an Internet connection per village. The computer is usually located in the Village Committee office or the Farmers' Book Houses. Some 400,000 village-level grassroot service centers have been set up by the project.

Digital Library Promotion Project
The Digital Library Promotion Project (DLPP) was launched in 2010 as a joint initiative of the Ministries of Culture and Finance. It aims to create a nationwide digital library with an extensive collection of electronic holdings that can be accessed from any public library as well as via the public Internet and mobile phones.[4] The platform is being gradually refined and extended throughout the country with a target completion date of 2015.[5]

Use of public libraries
By the end of 2011, there were 2,951 public libraries at the county level and above. All provincial, regional, and municipal centers have at least one library and 90 percent of county seats had a public library.

Users
Public libraries are free and available to everyone. In recent years, the number of library users has grown steadily. In 2011, the number of users (at the county level and above) with library cards amounted to 22 million (figure 5.2, left). This represents only about 1.6 percent of the total population. However, the data do not include users of libraries and library-like facilities (for example, Farmers' Book Houses, and so on) at the village level.

Types of Library Services and Their Use
Public library services vary with the size and level of the library. In general, the bigger and higher the level of the library is, the richer are its services, the greater its number of users, and the higher the level of funding. Public libraries at all levels provide traditional services such as lending, reference and consultation, information retrieval, and so on. In line with technological developments, public libraries have begun to offer additional services including those which enable

Information and Communications in the Chinese Countryside
http://dx.doi.org/10.1596/978-1-4648-0204-1

Figure 5.2 Public Library Cards and Distribution by Jurisdiction

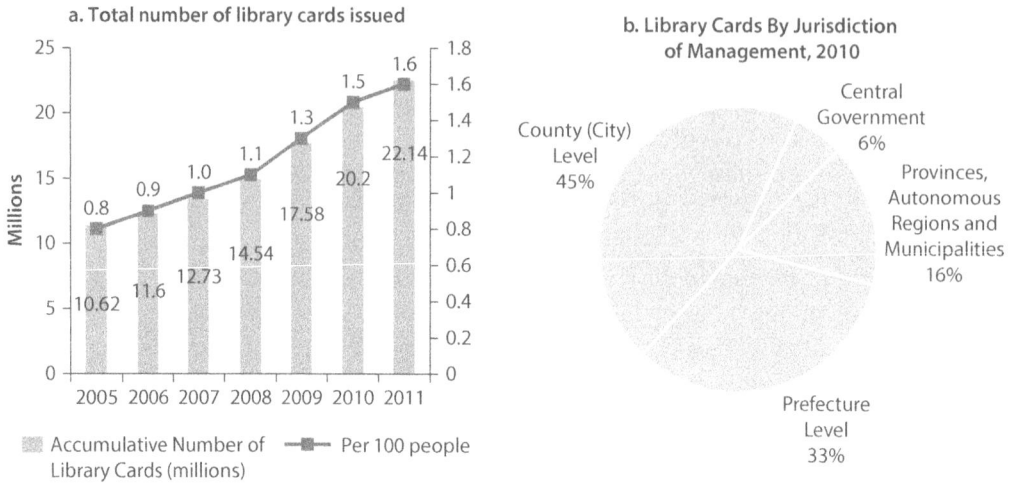

a. Total number of library cards issued

b. Library Cards By Jurisdiction of Management, 2010

County (City) Level 45%

Central Government 6%

Provinces, Autonomous Regions and Municipalities 16%

Prefecture Level 33%

Accumulative Number of Library Cards (millions) —■— Per 100 people

Source: Statistical Yearbook of China Culture and Heritage, and China National Bureau of Statistics.

Figure 5.3 Computers and Terminals in Electronic Media Reading Rooms in Public Libraries

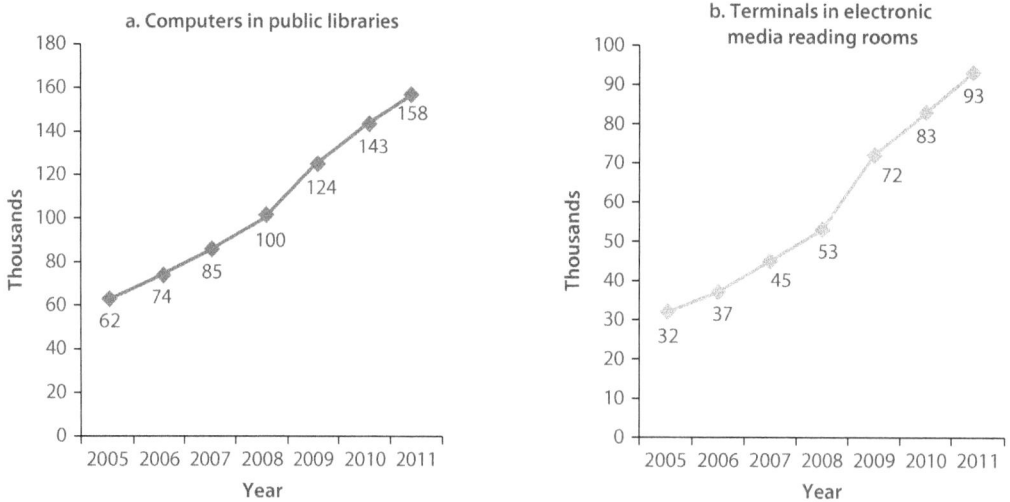

a. Computers in public libraries

b. Terminals in electronic media reading rooms

Source: National Bureau of Statistics.

access to the Internet and electronic content. The overall effect is a steady increase in the number of visits to public libraries.

Internet-Based Public Library Services

Chinese public libraries have set up electronic reading rooms to provide access to the Internet and electronic resources. The period from 2005 to 2011 saw rapid growth in the number of computers and electronic reading room terminals (see figure 5.3).

Rural libraries depend on local authorities for funding. National projects are only able to provide funds to establish, for example, a Farmers' Book House, but not to cover its operating costs. In some wealthier areas, such as Jiaxing in Zhejiang province, local laws have required the local government to provide operating budget for Comprehensive Cultural Stations and Farmers' House reading rooms, where they are branches of the county main library.

Library Survey

In order to better understand the operations of Chinese public libraries, particularly in rural areas, a questionnaire was designed and piloted. The questionnaires were then sent to library authorities between March and May 2012. The survey consisted of three questionnaires, each targeted at different levels of the public library system:

- Level 1: provincial and regional libraries
- Level 2: county libraries
- Level 3: township and village libraries.

Response rates varied markedly between individual questions, ranging from 98 percent to 5 percent or less (for example, questions regarding finances at Level 3). A variety of explanations may be relevant including: lack of understanding of terminology, perhaps particularly at Level 3 in rural areas; caution and other sensitivities regarding some of the information requested; and inability to supply some it. There is no guarantee that answers have not been exaggerated or understated. Nevertheless, the sample data gained for most questions appear sufficient to gain a reasonably reliable picture.

Key results of the survey are summarized here. They focus mainly on the findings at Level 3 (village/township). A fuller presentation is available in the *China Library Landscape Study* (Zhang and Davies 2013).

Almost 90 percent (87 percent) of Level 3 respondents were located in a rural area. Some four-fifths of libraries responding occupied a single service point (that is, only providing library services) (93 percent in Guizhou). The average (mean) distance of each branch from the nearest urban center was 25 kilometers. A third of library branches were located 20–50 kilometers from an urban center and 11 percent farther away than that.

Four-fifths of Level 3 respondents were participating in national government projects, 61 percent in local government projects, and 5 percent in projects with private/nongovernmental organization (NGO) support. Shandong is notable for the high proportion of local projects (85 percent).

While national statistics are not available and it is not possible to calculate directly from the survey data the total number of members of Level 3 libraries in the surveyed provinces, average per capita ratios have been calculated from the survey data, based on estimates of population served provided by the

Table 5.1 Average Membership Per Capita by Province and Library Level
percent

	3 provinces	*Guizhou*	*Jilin*	*Shandong*
Level 1 (provincial/regional)	1.3	1.2	1.6	1.2
Level 2 (county/township)	1.5	0.9	1.4	1.7
Level 3 (village)	8	2.6	9.9	6.1

Source: World Bank.
Note: Calculated using average population served figures provided by responding libraries.

libraries themselves (see table 5.1). These indicate that the percentage of the population using library facilities in rural areas is somewhat higher than that in the largely urban areas covered by the national data. In Jilin, for example, the figure is almost 10 percent. These data may support the suggestion of the utility of libraries in these areas (partly because of a scarcity of competing attractions).

On average, library membership in villages is split 56:44 male: female. Working age adults are the largest users of Level 1 and Level 2 libraries accounting for slightly less than a third of the members. Other members are split about evenly between children, students, and retired persons.

At Level 3, 14 percent of responding libraries provided Internet access. Less than 1 percent had their catalogue online, while 13 percent reported having some form of local information on the web. Book lending (95 percent), children's libraries (60 percent), and loans of nonbook items (30 percent) were the most common services at this level. Only 9 percent reported providing a reference information service, 17 percent had study space and fewer than 10 percent provided any kind of training.

Responses indicated significant differences in the popularity of particular services that libraries offer between survey levels 1, 2, and 3, although they generally involve both learning and recreational interests. At the village level, there was a low response rate to the question (no libraries in Jilin province responded to those questions). The impact of CIRSP in village libraries may therefore be fairly restricted both for reasons of connectivity and because of lack of interest in the content available.

By frequency of answer, the following content types were the most popular: (1) entertainment (films and TV plays); (2) web chat; (3) games; (4) e-books, e-journals, or e-news; and (5) law and agricultural practice.

About half of respondents indicated that their library or reading room was open for more than six hours a day. Observational evidence from fieldwork carried out in this study indicates a position somewhat less favorable than this.

At Level 3, 55 percent of libraries could not maintain existing technology with their own staff. This figure ranged from 79 percent in Guizhou to

30 percent in Shandong. Sixty-eight percent reported having access to other technical support (44 percent in Guizhou; 88 percent in Shandong). Among respondents, between 17 and 21 percent reported being able to introduce new technology through their staff in the past three years (the average was much higher in Shandong than elsewhere). Public Internet access, a website, and digitization technology—presumably a scanner—were almost equally the most common technologies reported. Over 70 percent of respondents did not answer this question.

Sixty-two percent of responding libraries connected to the Internet through Asymmetric Digital Subscriber Line (ADSL). There were few cases of optical fiber connection. Internet connection problems were common: For 14 percent of respondents they occurred more than once per day.

Less than 5 percent of responding libraries did not have a computer with Internet access. The rest had between one and five computers at Level 3 libraries. However, only 29 percent of responding libraries said that they provided Internet access to users and of these 23 percent made it freely available to anyone (12 percent in Jilin; 57 percent in Shandong). The reasons for this apparent anomaly requires further investigation and may relate to a lack of funds to maintain Internet connections, where computers have been provided through projects.

Fifty-seven percent of libraries at Level 3 report that computers were in use more than 50 percent of the time, although indications from fieldwork observations suggest that the figure is considerably lower.

At Level 3, 62 percent of respondents (nonresponse rate was 70 percent) spent less than CNY 500 ($90) in 2010. CIRSP was the main supplier of IT equipment.

The information and communications technology (ICT) demand survey carried out in rural areas of the three provinces also included several questions related to the availability and use of libraries.

- Availability of library in village: Sixty percent of the villages surveyed were reported to have a reading room or library (figure 5.4, top left).
- Availability of libraries providing public Internet access: A fifth of libraries reportedly offered public Internet access (figure 5.4, top right). Despite over half of villages in Shandong province having a library, none of them reportedly offered Internet access.
- Availability of library in school: Sixty percent of schools reported having a library or reading room (figure 5.4, bottom left). This ranged from around two-fifths in Jilin to four-fifths in Shandong.
- Public library visits: Six percent of household members interviewed reported visiting a public library in the past year (figure 5.4, bottom right). This ranged from 3 percent in Jilin province to 12 percent in Guizhou.

Information and Communications in the Chinese Countryside
http://dx.doi.org/10.1596/978-1-4648-0204-1

Figure 5.4 Availability and Visits to Libraries in Rural Areas of Three Chinese Provinces

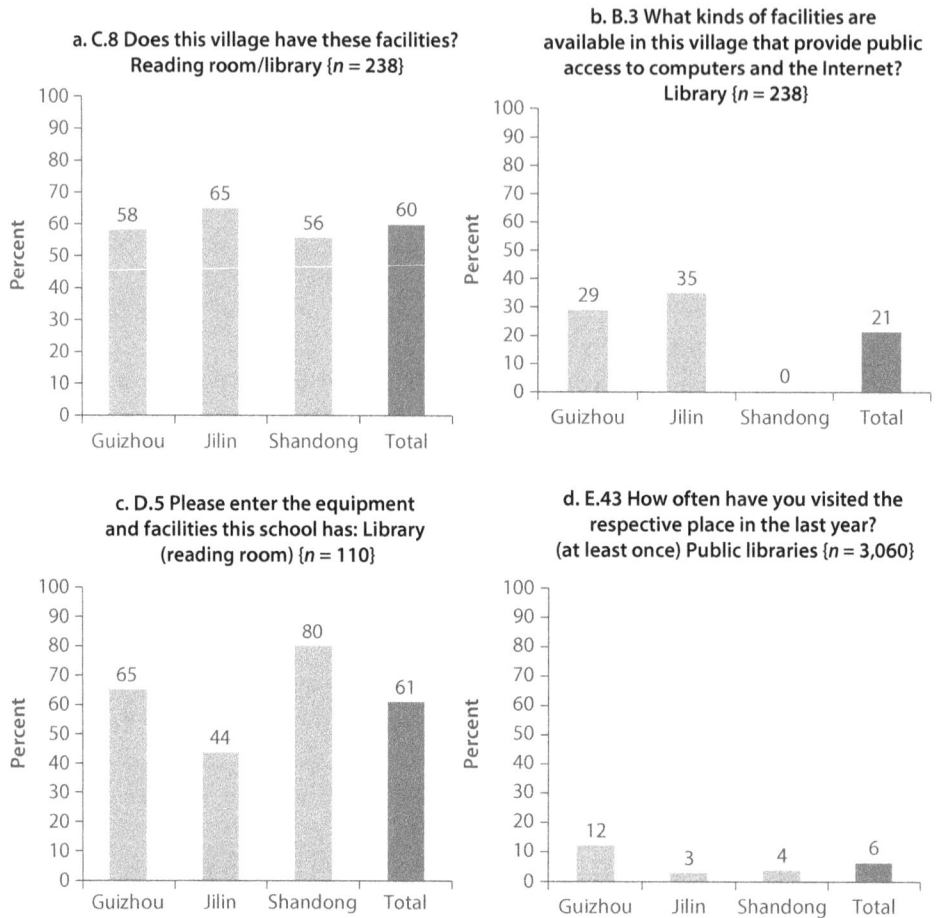

a. C.8 Does this village have these facilities?
Reading room/library {n = 238}

b. B.3 What kinds of facilities are available in this village that provide public access to computers and the Internet?
Library {n = 238}

c. D.5 Please enter the equipment and facilities this school has: Library (reading room) {n = 110}

d. E.43 How often have you visited the respective place in the last year?
(at least once) Public libraries {n = 3,060}

Source: State Information Center 2011.

Observations

China has made significant investments in the development of public libraries and library-like services. This has led to the establishment of several world-class libraries at the provincial and regional level that operate high standards of service and innovation (for example Hangzhou, Shanghai). Libraries in urban areas (that is, county and municipal level) are also pervasively established. There are however significant disparities, as in other aspects of social and economic life, between levels of provision and use in the eastern part of China and the rest of the country and between individual provinces. Only a small proportion of the population are public library members, although the number of registrations has grown consistently in recent years.

Several national projects have been established to provide library and information services in rural areas. Among these, the Public Electronic Reading Room Construction Plan covering villages throughout China is potentially of high

Information and Communications in the Chinese Countryside
http://dx.doi.org/10.1596/978-1-4648-0204-1

significance, although it is still too early to assess its impact. There is a relatively high level of participation by existing libraries in these projects, together with projects generated by local authorities—and a high level of dependence upon them for support.

The result of these initiatives has been to establish a widespread infrastructure of rural library-style services. Constraints and challenges affecting their use include: lack of permanent staff; lack of ICT knowledge and service management skills among staff; absence of regular operational budget and consequently high dependence for finance on national projects; and lack of useful space and functioning equipment, including ICTs.

As a result, these services are relatively lightly used. The effectiveness of their functioning is restricted by a number of key factors including weak access to operational funding beyond the project stage; consequent difficulty in recruiting and paying staff adequately trained in service management and delivery; limited ability to keep facilities open; and uneven network connectivity and equipment maintenance problems. The absence of effective administrative networks for libraries, which could enable larger libraries in urban areas to support and act as hubs for rural libraries in a region, is a further impediment to effective delivery. Despite the above, the data obtained in this study suggest a more intensive use in rural areas than in more mainstream urban library settings.

Notes

1. Examples of initiatives to transform libraries and advocate their ICT role include the Global Libraries program (http://www.gatesfoundation.org/libraries/Pages/global-libraries.aspx), and Beyond Access (http://www.beyondaccesscampaign.org/wp-content/uploads/2012/03/Beyond-Access-Libraries-MDGs-Small.pdf).

2. For example, see Qiu 2012.

3. Ministry of Culture, Ministry of Finance. Notice on the issuance of *"Public Electronic Reading Room Construction Plan" Implements Scheme*, 2012.

4. National Library of China (NLC). 2012. "National Library of China Annual Report to CDNL 2012." http://www.cdnl.info/2012/pdf/CDNL_Annual_REPORT_China.pdf.

5. http://www.library.sh.cn/Web/news/20111010/n57931640.html

CHAPTER 6

Impact Assessment

Monitoring the impact of rural informatization would provide the evidence to help determine which strategies are successful, which might need to be improved, and which seem to have little effect. However, rigorous monitoring and evaluation of rural information and communications technology (ICT) interventions has been largely absent in China. Therefore, the Chinese government encouraged an impact assessment framework as part of the project. The approach taken was to study existing ICT interventions in rural areas in terms of broad social and economic benefits.[1] Three interventions were selected for analysis (China Internet Network Information Center [CNNIC] 2013b):

1. Multipurpose telecenters in Guizhou province
2. Comprehensive information service platform in Jilin province
3. E-commerce platform in Shandong province.

Though this approach falls short of a more rigorous impact evaluation, it does shed light on several results from the interventions to accompany other anecdotal evidence of ICT effects in rural Chinese areas (see box 6.1).

Box 6.1 Use of ICT to Raise Incomes in Rural China

ICT can help increase rural incomes in various ways: (1) electronic access to agricultural information about weather, crop techniques, diseases, and prices helps farmers become more productive by reducing costs and losses; (2) advertising rural products over the Internet through e-commerce opens up new markets that can add to incomes; and (3) ICT-enabled work platforms offer new earning opportunities for low-skilled labor.

The impact evaluation study for this project found that the creation of **agricultural information** and its dissemination via mobile phones, the Internet, and radio and television programs is helping to raise farmers' incomes.

The use of **e-commerce** is being adopted by pioneering rural entrepreneurs who are starting to sell agricultural products, local specialties, and popular commodities over the

box continues next page

Box 6.1   Use of ICT to Raise Incomes in Rural China *(continued)*

Internet. Some have advanced from peddling agricultural products to developing supply chains including contracting local agents to process, sell, and distribute their products (CNNIC 2012).

Another promising area for raising rural incomes is the emergence of **ICT-enabled contract platforms** aimed at individual workers (Rossotto, Kuek, and Paradi-Guilford 2012). The range of tasks range from low-skilled data entry-like labor ("microwork") to more sophisticated computer programming. Individuals bid to perform the task and are paid through online payment systems. A report commissioned for this project (CNNIC 2013a) found that international mainstream microwork platforms have not been localized for China and thus do not have many users. Although somewhat similar platforms exist in China, they are in an embryonic stage and mostly geared to those with higher skill levels. The report recommends several steps for fostering microwork in China. This includes promoting the concept among Chinese enterprises; upgrading domestic crowdsourcing sites; training the rural population in using microwork platforms; and developing intermediation services through agents interfacing with companies and subcontracting simple, repetitive labor to rural citizens.

Description of Interventions

Guizhou

In 2007, Guizhou Province started to establish multipurpose telecenters (MTs) in rural areas on a trial basis. Between 2008 and 2011, MTs were established in ten villages each year. The program was accelerated in 2012, when MTs were established in 30 villages. So far the province has established MTs in 77 villages, with the largest number in areas populated by ethnic minorities.

As the name implies, the MTs serve multiple uses. They feature an open space containing a small reading room, a seating area with a large video monitor, and desks with computers and Internet access (see figure 2.4). There are 10–15 personal computers (PCs) per MT and a minimum of 10 megabytes per second download Internet speed delivered over fiber-optic cable. Specialized training is provided for the managing staff of the MTs and basic ICT training for village leaders, rural brokers, and members of farmer cooperatives.

In the province's view, if informatization is to develop in rural areas, villagers need to be given access to computers and the Internet, even if they initially tend to use it for social networking and entertainment. They will then gradually realize the importance of the Internet for more meaningful socioeconomic benefits. Therefore, one of the main purposes of MTs is to give villagers access to the Internet and, with proper guidance such as various kinds of training, familiarize them with it and create the need to use it. The establishment of MTs also provides rural areas with well-equipped and convenient public places for access to the Internet, which can help village governments attract investment from enterprises elsewhere and develop ICT-enabled cottage industries.

The establishment of village-level MTs for farmers is led by the Meteorological Bureau of Guizhou Province and supported by the Communications Administration, the Department of Culture, the Department of Commerce, China Telecom, and China Unicom.

Jilin

12582 is China Mobile's number for its nationwide agricultural information service. Users dial 12582 from their mobile phone to receive agricultural information. The service includes extended features beyond basic text messaging and is also available from websites. In Jilin Province, the 12582 platform has been jointly developed by the Agriculture Commission and China Mobile Jilin (China Mobile Group's operating company in the province). Staffed by over 600 agricultural experts at the provincial, municipal, and county levels, it provides farmers with access to market information, weather service, experts' response to questions, information about plant/animal cultivation, plant disease and pest control, and scientific fertilization methods. The objective is to provide farmers with a high-quality, well-targeted personalized service in order to help achieve increases in production and incomes.

The province's 12582 Comprehensive Information Service Platform was launched in March 2008 and consists of two parts, one via text messages and the other via voice. Clients of China Mobile Jilin who have subscribed to the 12582 service are entitled to receive group text messages, interactive services, and radio programs over their mobile phone.

The 12582 service has been popularized through promotional efforts in collaboration with media outlets including television, radio, and newspapers. The 12582 service is available to all subscribers of China Mobile Jilin. The service does not require significant investment since mobile coverage is widely available throughout the province. Given that the availability of computers and Internet access in rural areas is limited, it is highly practical to provide this service through mobile phones, the possession of which is common in rural areas. According to the demand survey results, 92 percent of rural inhabitants in the province owned a mobile phone in 2011.

Application service providers collect original information and provide them to China Mobile Jilin. The latter reprocesses and formats the content and integrates them into the final product. China Mobile Jilin is mainly responsible for the development and promotion of the program and the provision of funds. The Agriculture Commission, which administers the program, is responsible for its execution. Day-to-day operation is carried out by Jilin Rural Economic Information Center. China Mobile Jilin provides the Agriculture Commission with a certain amount of funds each year to operate the program. The annual operating expenditure is about CNY 10 ($1.6) million. The funds for the 12582 service mainly come from three sources: financial support by Jilin Province, China Mobile's investment, and business-raised funds. Investments over the years have totaled about CNY 50 ($7.9) million. At present, the platform is operated as a nonprofit service for farmers. In the future, it is expected to develop a level of self-sufficiency.

Shandong

Over the years, various parts of the Chinese government, such as the Ministry of Agriculture, the Ministry of Science and Technology, the Ministry of Industry and Information Technology, and the Meteorological Administration, have made efforts to promote the development of e-commerce in rural areas. Local governments at various levels have supported these efforts or acted on their own. As a result, a network of organizations and support has come into existence for the growth of rural e-commerce. It is in such an environment that rural e-commerce in Shandong Province has gradually developed. Rural self-employed individuals, agricultural enterprises, and agricultural cooperatives have joined various types of e-commerce platforms. The evaluation examined the specific use of the Shandong Agricultural Development e-commerce platform.

Findings

The following findings are drawn from the evaluation of rural informatization interventions in the provinces of Guizhou, Jilin, and Shandong:

These programs **have benefited those who use them** in various ways. Well over half the users reported monetary benefits from productivity gains due to learning better production techniques, buying agricultural inputs at a cheaper price, and selling farm products through new sales channels (see figure 6.1, a). Nonmonetary benefits include acquiring Internet-surfing skills and thereby learning more about the outside world and broadening one's horizon (see figure 6.1, b). The acquisition of new ideas, concepts, and skills can help to change the "small farmer mentality" and modernize rural Chinese life.

Figure 6.1 Financial and Nonfinancial Benefits from Rural ICT Interventions

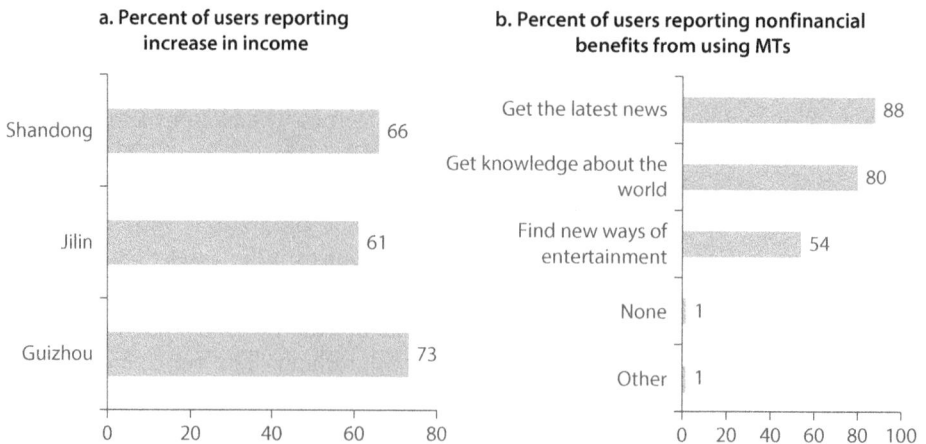

Source: World Bank.
Note: ICT = information and communications technology; MT = multipurpose telecenter.

The role of information differs by the nature of the intervention. In Guizhou and Jilin, the main monetary benefit resulted from efficiency gains through better production techniques. In Shandong, the main benefit came from finding new channels to increase product sales. The nature and audience of the intervention affects these findings; in Shandong the program exclusively focused on e-commerce and with a specific group of beneficiaries. In contrast, the interventions in Guizhou and Jilin were more general and required users to be more proactive to post marketing information. In addition the mobile phone orientation of the Jilin program makes it more difficult to develop extensive marketing information. In the case of Shandong, the impact of information was the potential for market transfer in the supply chain, whereas in the latter case information enhanced arbitrage through efficiency gains.

The programs are organized according to a **public-private partnership (PPP) model.** Cooperation between government agencies and telecommunication operators with obligations for universal service can guarantee their launch and initial operation. Users reported problems with lack of useful content, irregular operating hours or shortage of staff, symptoms of a lack of ongoing operational support. It may be necessary to introduce specialized partners for the day-to-day operation of the programs and to provide a more personalized and effective service targeted at rural needs. The government and telecom operators would offer assistance and favorable treatment in policy, equipment, and operational funding to ensure the pro bono nature of programs. The PPP model is also relevant for content development where the experience in Shandong province revealed that frequent use was made of Baidu, Alibaba, and HC360, highly experienced in Internet operation. In agricultural informatization programs, content (agriculture) and tools (information technology [IT]) are both essential and should be closely combined. Full use should be made of the knowledge-related authority of agricultural institutions and the service-related effectiveness of Internet companies so that they could jointly launch reliable and user-friendly products and services for agricultural informatization.

Programs differ in communication channels affecting user initiative and pro-activeness. For the multipurpose rural telecenters in Guizhou, once-a-week use is most common among the users (see figure 6.2, a). For the 12316/12582 service in Jilin, there are significant differences in the frequency of the use of specific services: Most of the users dial the hotline less than once a month; most of them receive and read text messages and tune in to radio and TV programs every day; but the website is rarely visited (see figure 6.2, b). Text messaging is a passive service and radio and TV programs are also largely passive. In comparison, a personal visit to the telecenter requires a high level of initiative; dialing the hotline costs money in addition to requiring strong initiative; visiting the website requires, in addition to strong initiative, access to computers and the Internet as well as IT-related

Figure 6.2 Frequency of Use of Rural ICT Interventions

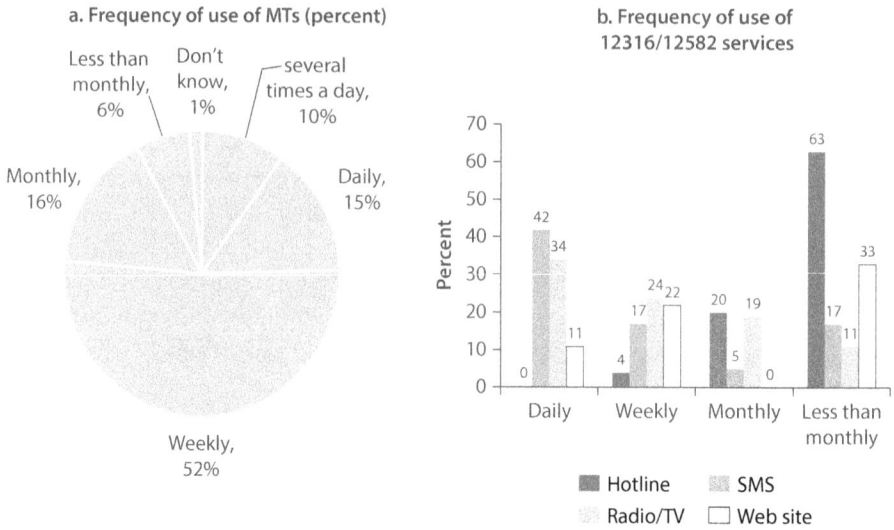

a. Frequency of use of MTs (percent)

b. Frequency of use of 12316/12582 services

Source: World Bank.
Note: ICT = information and communications technology; MT = multipurpose telecenter; SMS = Short Message Service.

knowledge. Users remain passive in their utilization of services owing to financial, technical, and skill restrictions as well as personal reluctance.

The studies show that the **users of rural informatization programs tend to be an exclusive group**. They are predominantly male and farmers; the unemployed, housewives, and the retired hardly use them (see figure 6.3). It is hoped that early adopters will influence others so that they are motivated to join the programs. However, a key obstacle is the agricultural focus of the programs; they lack applications in socioeconomic applications that support rural people's daily lives as well as in skill development which encourage rural people to diversify their incomes (such as ICT-enabled microwork, and so on).

Villagers tend to lack IT skills. This is a basic national condition to bear in mind when evaluating most Chinese rural informatization programs. For instance, the primary reason for not using MTs in Guizhou is the lack of IT skills (figure 6.4, a) and e-commerce users in Shandong mainly use the Internet to post messages but rarely use online transactions. In Jilin, half the users of the mobile phone platform are not able to use computers or the Internet (see figure 6.4, b). The organizers of the programs have launched various IT skills training programs. It is necessary to step up efforts in this regard in order to modernize Chinese rural areas. Rural informatization can only be successful as long as they can dramatically improve digital literacy and enable the majority of rural inhabitants to interact with the outside world with more ease and convenience.

Figure 6.3 Characteristics of Users of Rural ICT Interventions

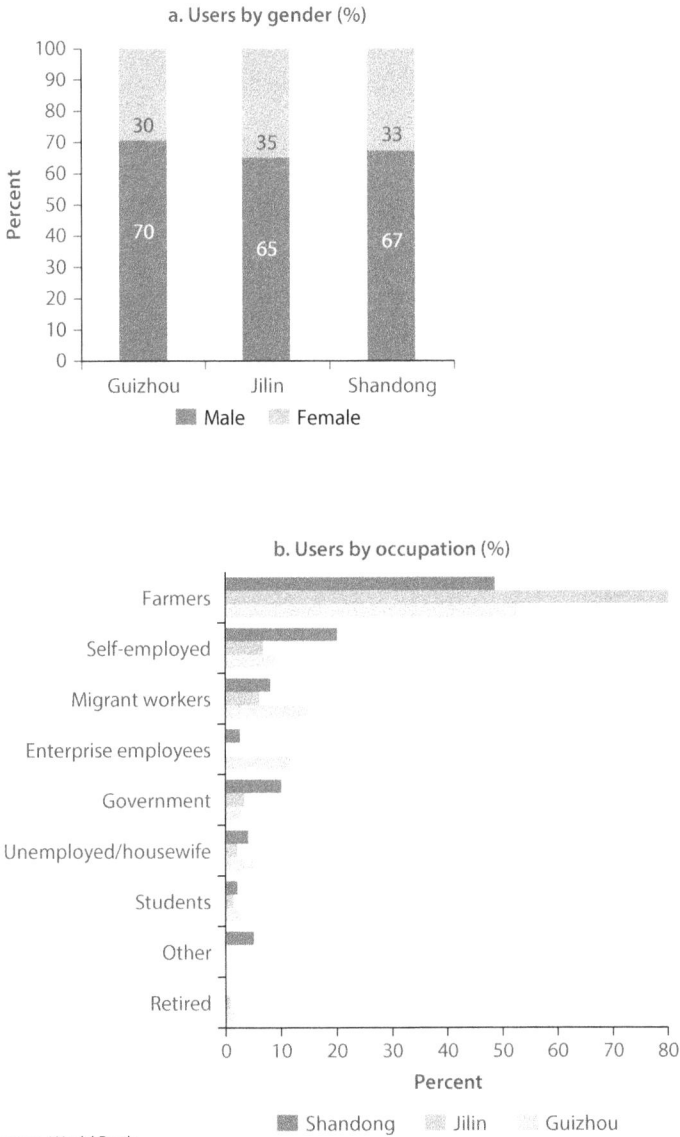

a. Users by gender (%)

b. Users by occupation (%)

Source: World Bank.

Figure 6.4 Reasons for Not Using MTs and ICT Skills Among Mobile Platform Users

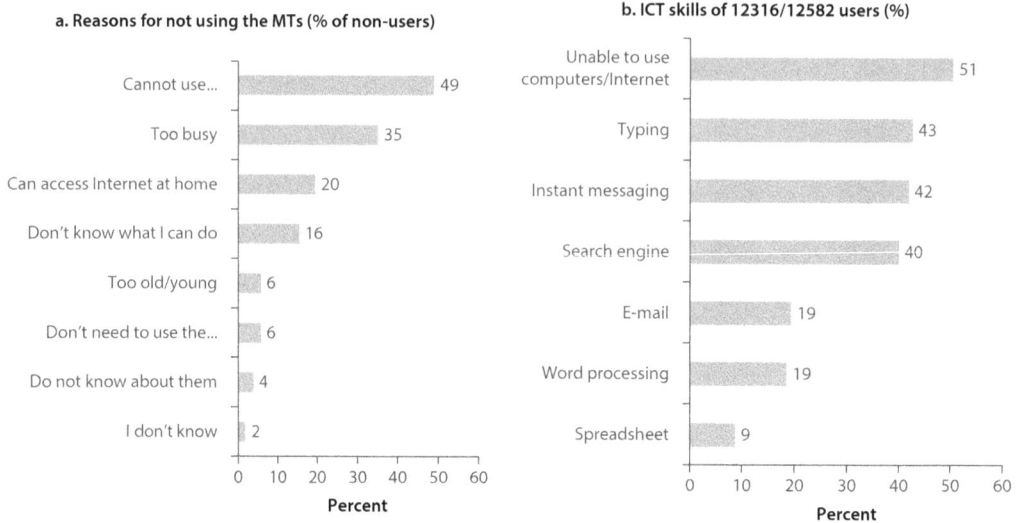

a. Reasons for not using the MTs (% of non-users)

Reason	Percent
Cannot use...	49
Too busy	35
Can access Internet at home	20
Don't know what I can do	16
Too old/young	6
Don't need to use the...	6
Do not know about them	4
I don't know	2

b. ICT skills of 12316/12582 users (%)

Skill	Percent
Unable to use computers/Internet	51
Typing	43
Instant messaging	42
Search engine	40
E-mail	19
Word processing	19
Spreadsheet	9

Percent

Source: World Bank.
Note: ICT = information and communications technology; MT = multipurpose telecenter.

Observations

Evaluations about the outcomes and impacts of rural ICT interventions are largely absent, making it difficult to determine which projects are successful and worth scaling up. This includes analysis of the specific modes in which interventions are delivered and, in the case of agricultural initiatives, understanding how the use of technology relates to inputs and outputs and their impacts on rural welfare. The role of information for arbitrage and market power is also worth examining as it could have wider implications on agricultural supply chains.

The project carried out a limited evaluation of an intervention in each of the three provinces. The main findings include the following:

- The purpose of the information varied significantly according to the nature of the intervention—telecenter, mobile information service, or e-commerce
- The programs helped those who use them with well over half the participants reporting productivity gains as well as nonmonetary benefits from ICT access
- The interventions used different communication channels, which required different levels of user initiative
- Users of rural informatization programs tend to be an exclusive group
- Most Chinese villagers lack IT skills.

In the future, the Chinese government might want to consider integrating monitoring and evaluation as a fundamental component of any rural informatization intervention. This could be carried out in partnership with local universities and research groups. Some impact evaluations have been carried out in China mainly in other sectors but there are also a few on ICT interventions in education. These, along with examples of international evaluations, are explored in a methodological report commissioned for the project (Cai 2013) and can serve as background for developing more rigorous evaluations of outcomes from rural ICT initiatives.

Note

1. Originally the plan was to leverage the ICT demand survey to derive a base dataset and then to reinterview a subset of respondents following a target intervention. This proved difficult since a comprehensive impact evaluation would have required the inclusion of a considerable number of extra questions in the survey. This was not possible, given that the demand survey was already at the limit of a tolerable timeframe for respondent completion. It also proved difficult to identify and align interventions with villages that needed to be randomly selected according to the statistical guidelines.

CHAPTER 7

Conclusions and Recommendations

China has made notable progress in extending information and communications technology (ICT) infrastructure to rural areas. By 2012, all of the nation's administrative villages were connected to the telephone network, broadband connectivity was available in 88 percent of administrative villages, and there were 156 million rural Internet users. Extensive agricultural content has been created through national and provincial initiatives. The results from this project's studies and surveys illustrate the positive results of rural informatization:

- The ICT demand survey found high use of mobile phones, available in 95 percent of rural households
- The library landscape study found national projects have established a widespread infrastructure of rural library-style services
- The impact assessment found that well over half the respondents reported monetary benefits from using selected rural ICT interventions

Nonetheless, there remain sharp differences in Internet use between urban and rural areas in China. While three-fifths of urban Chinese use the Internet, less than a quarter of the rural population do so—and the gap is growing.

China faces a number of challenges in its quest to improve rural ICT access through sustainable programs:

- **Facilities and services underused.** The Chinese government has made significant investments in rural ICT infrastructure, equipment, content, and public access facilities. However, this increase in supply has not been matched by a corresponding rise in demand. Less than 5 percent of respondents in the demand survey reported using the Internet in their own village; some may not be aware that such a facility even exists. Further, the structure of Chinese government ICT spending is oriented towards infrastructure and equipment rather than operating expenditure. The symptoms of inadequate operational support for rural ICT interventions are reflected in feedback collected from the surveys where respondents reported a lack of staff, irregular operating hours, and limited training.

- **Women and particularly senior citizens largely excluded.** Results from the surveys indicate comparatively low ICT usage by women and older rural inhabitants. One reason is an element of intimidation with many public Internet facilities oriented towards entertainment and dominated by young men. Another factor is that most applications and content developed for rural areas consist of agricultural information aimed at farmers. There is a shortage of content with an explicit focus on women and older people, who because of large-scale rural-urban migration, increasingly account for a significant proportion of the population in Chinese villages. Application development needs to be more inclusive, moving beyond agriculture to other areas affecting rural livelihoods, such as government services, health, and employment. For example, the use of ICT to generate contractual work opportunities has not been explored despite its demonstrated potential in other countries.
- **Scarce training possibilities.** According to the demand survey, the main barrier to Internet take-up is lack of skills. Yet few rural public Internet centers offer regular training opportunities, only a quarter of schools allow the community to use their computer labs, and rural libraries do not have the staff or facilities. There are increasingly limited opportunities for learning from family members since those that most likely would have some ICT skills are migrating to urban areas. Early ICT adopters are expected to impart their skills to other villagers but so far there is little evidence that this is the case.
- **Significant duplication.** Similar telecenter, e-commerce, and mobile phone information systems developed by each province result in considerable duplication of resources. This makes it difficult to achieve economies of scale and to identify best practice that could be replicated elsewhere.
- **Absence of business models for long-term sustainability.** There is concern about how interventions will be sustained after they have been launched. Few studies exist on the usage, impact, and cost benefit of Chinese rural ICT programs. This makes it difficult to determine what works and what does not, what needs to be scaled up, and what should be terminated. Furthermore, existing infrastructure such as public libraries are not effectively supported for widening digital access.

Building on the experience of rural ICT initiatives implemented to date and incorporating international best practice, the government is advised to consider new models for rural informatization programs to achieve higher impact with economies of scale and lower investment costs. Recommendations include the following:

- **A more coordinated approach.** Greater coordination and information exchange among the provinces would help to minimize duplicate efforts. Creating an advisory group consisting of ministries involved in rural informatization could be useful in this regard. Such a group could organize workshops on rural informatization with the aim of scaling up and standardizing common applications and services. An example is Australia, where the

Department of Communications managed the Digital Regions Initiative working with state, territory, and local governments to use communications networks for the delivery of education, health, and emergency services to rural areas.[1] Greater cooperation with the private sector should be encouraged, particularly Internet companies, to leverage their expertise in user-friendly interfaces, e-commerce platforms, and social networking applications.

- **Demand stimulation.** Demand can be increased through attractive and inclusive services and applications that enrich the quality of life for all rural citizens and widespread digital training. Applications need to move beyond purely agriculture and be more tailored to rural needs such as government services for land registration, health, and pensions as well as targeted online services for small and medium enterprises (SMEs), women, children, and older people. The development of a "killer app" for these services might help to popularize ICT use in rural areas. The technical and consultation skills of personnel operating informatization programs should be upgraded so they can provide comprehensive advice on how ICTs can be used to solve the full spectrum of rural residents' needs. Computer training needs to be upgraded particularly for underserved groups in rural areas (that is, older people, women, and children), and schools made available after hours to provide such training to the local community. Libraries could extend their functionality by offering computer learning. Several countries offer good examples of targeted digital programs for students,[2] older people,[3] and women.[4] Digital literacy would not only strengthen the ability of rural inhabitants to use existing ICT applications, such as online agricultural information, but also develop expertise in ancillary areas to add value to their existing activities or gain new revenue-generating skills to supplement farming income. Ongoing outreach campaigns such as open days, competitions, and marketing at local gatherings should be developed to raise awareness of rural ICT programs. Successful impacts should also be demonstrated so that villagers can see the concrete benefits of ICT use.
- **Improve monitoring and evaluation.** Data on the results of rural ICT interventions should be collected, compiled, analyzed, and disseminated on a regular basis. One of the reasons for duplicate systems across provinces is the shortage of information regarding the outcomes of various programs. Users and nonusers of the various programs should be surveyed both ex ante and ex poste to generate credible evidence regarding impacts. The project collected research on impact evaluations in China and also developed a model that could be applied (Cai 2013). The Baltic countries have developed impact-monitoring tools including metrics for calculating the economic impact of ICT access and training.[5]
- **Stimulating innovation.** Telecenters and libraries could be leveraged to become hubs of village innovation through networking rural businesses and communities. For example in the United States, North Carolina has deployed Business and Technology Telecenters (BTTs) in order to provide small business support through provision of meeting space; training, advising, and

mentoring; office needs (Internet access, fax, photocopying); and market development through activities such as farmer's markets and art exhibitions.[6] One strategy is to develop linkages with international and domestic firms for ICT-enabled contract work. In that regard, brokers should be supported to provide intermediation services between rural workers and companies and to train villagers how to use microwork and freelance platforms (China Internet Network Information Center [CNNIC] 2013a). The use of e-commerce and social networking platforms by small business should be fostered in order to improve productivity and expand access to domestic and global markets.

- **Focus on sustainability.** China could adopt several practices to achieve greater long-term sustainability for rural ICT interventions:

  – It needs to make greater use of partnerships in day-to-day operations. This includes shifting the obligations of telecommunications operators from infrastructure deployment to operations. This has been successfully used in countries such as Malaysia, where telecommunication companies operate telecenters with finance from universal service funding and where the fund is also used to implement broadband in rural libraries.[7]

  – Libraries are logical partners for rural informatization with their community-facing infrastructure, the fact that these venues already exist, and whose mission is already to provide public access to information. Some 60 percent of villages surveyed in the three provinces reported having at least a reading room. They could be an effective instrument for providing public Internet access and ICT training, as has been the case in Latvia.[8] The government and librarians recognize the need to upgrade libraries and train librarians to "Library 2.0"[9] so that staff members are well versed in ICT and better able to meet the needs of clients (Zhang and Hao 2012). The government should give consideration to greater support to the already existing network of public libraries for them to further develop their public Internet access and training capabilities. It is likely this would be more cost effective than funding brand new interventions. The innovative model of urban libraries nurturing rural ones could also be applied to enhance sustainability.

  – Public-private partnerships could also be employed more broadly for content development, e-commerce applications and in other areas to leverage the expertise of China's Internet companies.

  – Income-generating services could also be incorporated to help defray costs. For example, business services such as printing, faxing, and scanning should be available at telecenters and libraries. In addition, transaction fees could be charged for value-added applications such as e-commerce posting and payments. In Bangladesh, entrepreneurs operate government e-services centers and they are encouraged to develop additional revenue-generating services.[10]

- **Complementarity of access devices and facilities.** ICT access from computers and mobile devices should complement each other.[11] Mobile phones are ideal for personalized, short sessions, and small information streams with the confidence that the user generally has their phone with them. Computers and tablets are suited for more intensive applications, extensive searching, and e-book reading. Some countries, such as Thailand, have recognized the growing spread of mobile devices by launching free Wi-Fi access.[12] Libraries and telecenters could extend their portfolio by more deeply integrating wireless support. For example, Wi-Fi could be offered at every library and telecenter through the simple addition of an inexpensive router; extending the Wi-Fi signal throughout the village could provide additional convenience. Libraries and telecenters should also be configured so that there is appropriate space for users with laptops or tablets. Telecenters could also exploit caretaker and user ICT skills by serving as mobile content development zones to create localized text alerts about weather conditions, important village announcements, and so on.

These holistic approaches, featuring delivery of appropriate information and entrepreneurial management and supported by clear measurement results, are likely to have wide economic benefits for rural communities, with the goal of integrating Chinese villages into modern society.

Notes

1. http://www.archive.dbcde.gov.au/2013/august/digital_regions_initiative.

2. Uruguay has provided free laptop computers to all primary school students and teachers in order to democratize ICTs and encourage spillover effect of children teaching parents (Trucano 2010).

3. The Australian government's Broadband for Seniors program provides free computer training and Internet access to the elderly at special kiosks. Volunteers offer hands-on training in a friendly environment with courses specifically designed for older people. See: http://www.dss.gov.au/our-responsibilities/seniors/programs-services/broadband-for-seniors.

4. The Philippine Digital Literacy Campaign for Women has provided computer training to over 10,000 women. See: http://www.philcecnet.ph/content/view/777.

5. Research conducted in Latvia and Lithuania measured the outcome of public access computing. For the methodology and main results, see Paberza and Rutkauskiene (2010).

6. Since 2001, the BTTs have created more than 1,900 jobs, leveraged more than US$13 million in outside funding and business revenue, served more than 26,500 business clients, and provided Internet access to more than 200,000 users. See: http://ncbroadband.gov/assistance/e-nc-business-technology-telecenters.

7. Malaysia has been building on its decadelong experience using regulatory tools to enhance rural ICT infrastructure and access. Since 2007 it has installed 251 Community Broadband Centers (CBCs) administered by the Malaysian Communications and Multimedia Commission (MCMC) with deployment and operating costs covered by the universal service fund. They are deployed in underserved areas of the country providing information access, e-services, training, and small enterprise development. The CBCs are managed by telecommunication operators and employ two skilled staff from the local community. See: MCMC. 2012. *2011*

Universal Service Provision Annual Report. http://www.skmm.gov.my/Resources/
Publications/Universal-Service-Provision-Annual-Reports.aspx.

8. The Latvian Trešais tēva dēls ("Father's third son") project, launched in 2006, has
transformed the country's libraries into advanced ICT venues. Computers and supporting software were installed in all 68 public libraries (including special computers for the blind at four libraries) and Internet access speeds were upgraded. In
addition, all libraries were Wi-Fi enabled. All librarians received training in computers, software, user support, and innovation; training in digital literacy was also provided to users. For those with low incomes, the availability of free services was
important; over a quarter of Latvian Internet users reported using libraries to go
online, of which 97 percent stated it was their main point of access. Within particular library services, the use of PCs and the Internet had the second best benefit-cost
ratio after exhibitions. See: "Library development project 'Father's third son,'"
http://www.projectdla.eu/dla/sites/default/files/LATVIA_e-democracy%20(3).pdf,
and Kultūras informācijas sistēmas. 2012. *Economic Value and Impact of Public
Libraries in Latvia.* http://www.kis.gov.lv/download/Economic%20value%20
and%20impact%20of%20public%20libraries%20in%20Latvia.pdf.

9. Library 2.0 means making library space (virtual and physical) more interactive, col-
laborative, and driven by community needs.

10. Union Information Service Centers (UISCs) have been installed in all 4,498 of the
country's unions, the lowest administrative division. The UISCs are operated by a
team of two entrepreneurs including at least one woman in order to attract female
users. The UISCs are essentially the main source of information services for
Bangladesh's rural population who constitute about three quarters of the population.
They offer services such as Internet access, e-mail, videoconferencing, downloading
government forms, scanning, printing, and digital photography processing. In addition,
the UISCs provide a growing number of value-added services such as mobile recharg-
es, money transfer, and access to the national e-content repository in the Bangla lan-
guage. After receiving free training, the entrepreneurs are responsible for operating
costs and charge a fee for services to ensure sustainability. Local governments provide
a room and the initial equipment. See: United Nations Development Programme
(UNDP). 2011. *Access to Information (A2I): Our Stories of Achievements.*

11. There is a growing body of research of the roles and benefits for different ICT devices and how they can complement each other. See: Walton and Donner (2012).

12. Thailand had over 250,000 free Wi-Fi hotspots in mid-2013 with plans to reach
400,000 by 2014. The hotspots are an initiative of the government who has encouraged the country's telecom operators to deploy the service. Service is limited to
around 20 minutes per session and up to two hours per day. See: Basu (2013).

Key Rural Informatization Initiatives by Central Government Ministries and Agencies

Ministry/agency	Key projects	Year(s)
Central Committee of the Communist Party of China (CCCPC)	Modern Distance Learning of National Party Cadres in Rural Areas Project	2003–06 Distance education is provided at 645,000 locations (2009)[a]
Ministry of Agriculture	Three in One (Telephone, Television, and Computer) Agriculture Information Services Project	2005
	Overall Framework of National Agriculture and Rural Informatization, 2007–2015, and pilot projects	2007–
Ministry of Commerce	Thousands of Villages and Townships Project	2005
	Xinfu Project (Commercial information services system for the countryside)	2006
	Home Appliance Subsidy Program for rural areas[b]	2007–13 Provided a subsidy for purchase of appliances including computers
Ministry of Culture	National Cultural Information Resources Sharing Project[c] (with Ministry of Finance)	2002–
	Comprehensive Culture Station Project	2006–10
Ministry of Education	Distance education project for rural schools[d] (with Ministry of Finance)	2003–07 Aimed at rural schools in central and western China with three models: 1. DVDs, DVD player, and TV set for 110,000 low-level primary schools 2. DVDs, DVD player, TV set, and satellite dish for 384,000 intermediate primary schools 3. Computer labs for 37,500 junior secondary schools. By 2007, the project had been implemented in 80% of schools.

table continues next page

Ministry/agency	Key projects	Year(s)
Ministry of Industry and Information Technology	Extend Telephone Coverage to Every Village Project[e]	2004–
		By November 2011, 100% of administrative villages and 94.5% of natural villages were connected to the fixed telephone network.
	Information to the Countryside	2009–
		Encourage operators to establish information stations and databases on agriculture.
Ministry of Science and Technology	State Agricultural Science and Technology Park Development Program	2007
	"Spark" Agricultural Science and Technology 110 Information Services Project	2005–10
State Administration of Radio, Film, and Television	Extend Broadcasting (TV and Radio) Coverage to Every Village Project	2006–10 (phase 2)
		Radio population coverage extended to 95.4% of the population by 2007 with television population coverage at 96.6%.[f] By 2011, all villages with at least 20 households and electricity had radio and television coverage; by 2015 coverage will be extended to villages with less than 20 households using direct broadcast satellite technology.[g]

Source: World Bank.
Note: DVD = digital versatile disc.
a. http://eg.china-embassy.org/eng/rdwt/P020110529735490640280.pdf.
b. http://english.peopledaily.com.cn/90778/8115206.html.
c. Wu Xiao (2012).
d. McQuaide (2009).
e. http://www.itu.int/wsis/stocktaking/scripts/documents.asp?project=1142247649.
f. http://www.sarft.gov.cn/articles/2008/04/30/20080430174159330771.html.
g. http://english.peopledaily.com.cn/90882/7589485.html.

ICT Demand Survey

Survey Design and Implementation

The World Bank project team created a questionnaire for the survey to discover more about the availability and awareness of, and attitudes towards, information and communications technologies (ICTs) in rural areas.[1] The questionnaire was based on similar surveys carried out in China as well as other countries. The questionnaire was structured into five sections revolving around different target groups:

A. **Public access point**: The current status of places where the public can go to use computers and the Internet was surveyed. This included interviews with the person responsible for the public access point in the respective villages; if there was no public access point in the village, then the public access point that rural households go to in another location was surveyed. A questionnaire was designed for the survey staff to enter information about the facility. This consisted of an equipment inventory and capacity of the facility (for example, number of computers, type of Internet connection (including speed), number of printers/scanners, and so on). Although the surveyors asked the manager of the facility to complete the initial questionnaire, the surveyors inspected the premises to ensure that the actual situation reflected the data provided on the questionnaire. This would include, for example, using a workstation by turning it on, trying to access Internet, printing, and so on. Interviewers were also asked to assess the overall situation of the public access point (for example, noise, location, cleanliness, and so on) and to take a photo of the establishment. The purpose of this part of the survey was to make an assessment of public access points including who uses them, the extent of usage and the equipment available. This helped inform analysis regarding the role and importance of public access points in the village.

B. **Community leaders**: This part of the survey consisted of interviews with community leaders, such as local government officials. They were queried regarding the status of public ICT facilities, including what other venues might be appropriate for providing ICT services. Community leaders were asked standard questions for their feedback. In addition, the interview with local community leaders revealed their impressions about attitudes, needs, awareness, and usefulness of ICTs for the village. The output from the interviews provided a community-level perspective about the adequacy of current access to ICTs, the possibility of widening the scope of localities providing ICT access, and the kinds of interventions needed.

C. **Infrastructure audit:** This part of the survey consisted of village infrastructure profiling. This tabulated ICT facilities available in the village including data such as the number of telephone lines, number of mobile subscribers, number of Internet subscribers, coverage of mobile network, and so on. The village "accountant" was interviewed to obtain this data.

D. **School principal:** This part of the survey was aimed at primary and secondary school principals regarding the status and availability of ICTs in educational institutions. It included questions such as the availability of ICTs in the school, whether they were available for students and the local community to use, ICT training and curriculum, and maintenance. These interviews allowed the state of ICT in the rural schools to be assessed.

E. **Household:** This part of the survey consisted of a structured questionnaire for the household segment in villages in order to determine (i) current access to information and communications; (ii) information sources in daily life; (iii) information needs (for example, content needs, service needs, training needs, and so on); (iv) perceptions of the value of ICT; and (v) usage patterns. The household interviews provided an assessment of the availability of ICTs, how they were used, what kind of information was deemed important and what problems rural householders faced in obtaining and using ICT.

Terms of Reference were prepared to carry out the survey. China Mainland Marketing Research (CMMR), founded in 1994 and previously the consulting department of the Chinese Bureau of Statistics, was selected. Their proposal outlined the approach for a statistically representative survey including the number of villages and households to be interviewed and the village selection process.

A pilot survey was conducted in Jilin province in April 2011 to test the questionnaire (see figure B.1). A workshop was held in Changchun, the capital of Jilin province, for the local interviewers and included training, background information about the project, and other technical details. The pilot was carried out in two villages (Gu Jia Ling Zi, and Bai Jia Ying Zi). In each village, the team met with community leaders including the village chief and school principal, and visited the nearest public access point (for example, Internet café). Ten households were interviewed in each village.

After the pilot survey in the villages, a review meeting was held in order to revise the questionnaires based on the field experience.

The full survey was carried out between September 26, 2011, and November 2, 2011. Prior to that, the sample villages were selected, questionnaires printed, and field staff trained. Ten counties were selected in each province. Within each county, 8 villages were selected. Then within each village a minimum of 10 households was

Figure B.1 Pilot Survey in Jilin Province

Source: World Bank.
Left: Members of the pilot survey team. Right: A local farmer being interviewed

selected. There was an oversampling of households to allow for invalid question-naires. The sample selection technique and other methodological issues are addressed below.

Questionnaires were audited on a daily basis to correct any errors and, if necessary, reinterview while interviewers were still in the village. Data were uploaded to the database on a regular basis to enhance the timeliness of the process. Postprocessing of the data was completed on November 20, 2011.

The final number of households, villages, schools, and public Internet facilities surveyed is shown in table B.1.

Discussions were held in Beijing in December 2011 between the World Bank, China Mainland Market Research Co., Ltd (CMMR), and State Information Center (SIC) to consider the results of the survey. SIC was charged with taking the lead on producing a report summarizing key findings of each section of the survey.

A draft of the demand survey results was prepared for expert group review. The *ICT Demand Survey Review Workshop* was held in Beijing on March 21, 2012. The workshop introduced the survey methodology and fieldwork and the main results. About 20 rural ICT experts from various ministries and provincial authorities attended and provided insightful comments. Input from the workshop was incorporated into the final report (SIC 2012).

Table B.1 Number of Households, Villages, Schools, and Public Internet Facilities Surveyed

	Jilin	Shandong	Guizhou	Total
Households surveyed	1,019	1,029	1,012	3,060
Schools surveyed	32	15	63	110
Villages surveyed	80	79	79	238
Average population	1,570	980	2,047	1,533
Public Internet facilities surveyed	79	80	48	207

Source: World Bank.

Sampling Method for Demand Survey

1. Sampling population: all qualified people in all administrative villages in the three provinces
2. Sampling rules:
   - Sampling of counties in the three provinces: Classify all counties in each province into layers by region/city and sample one county in each layer.
   - Sampling of villages in sampled counties: Sort all administrative villages in such counties and sample at random.
   - Sampling of family households: Sample in villages by direction, that is, respectively, sample a household in the east, south, west, north, and center of each village.
   - Sampling of family members: Pay a visit to a family member who knows the most about production and living conditions and family information and communication equipment in each household.
3. Sampling method:
   - First step: sampling counties in such provinces
   - Second step: sampling villages in counties sampled above
     - Sample an equal number of villages at random in the counties sampled from each province above.
     - Sample eight villages in each county.
4. Sampling of rural residents:
   - First step: sampling households in villages sampled above
     - Sample households by means of random start and equal distance to ensure that sample households are evenly distributed in villages.
     - Sampled households shall be the ones that have a continuous period of residence in a village for half a year or more.
   - Second step: sampling of family members in households sampled above
     - Pay a visit to the family member of each household to know the household's production and living condition and the use of family information and communication equipment.
     - Family members aged from 16 to 69.
5. Sampling of village leaders
   - First step: Sample village leaders who know the actual situation of their villages and have representative opinions.
   - Second step: If there are several qualified leaders in one village, the ones present at time of survey shall be visited first to improve work efficiency.
6. Sampling of elementary and secondary school leaders
   - First step: Sample all elementary and secondary schools in sampled villages
     - If there is no elementary or secondary school in a sampled village, it is unnecessary to complete this questionnaire.

- Second step: sample school leaders
  - Visit those persons who know the most about the overall situation and use of information and communication equipment of their schools.

7. Sampling of public access points
   - First step: If there is only one public access point in a village, such point will be visited.
     - If there are several public access points in a village, the public access point where villagers go most frequently will be visited.
   - Second step: if there is no public access point in a village
     - Pay a visit to a public access point (Internet café) where villagers go most often or the nearest to such village; such Internet cafés may be in a town or another village.
     - If there is no public access point in a village and village leaders and villagers have no idea of the nearest public access point, the questionnaire shall be abandoned, with the reason recorded.

Sample Size Calculation

Household Sample Size

According to the China Mainland Marketing Research (CMMR), the company that carried out the fieldwork, the survey project was mainly to focus on investigating the informatization status of rural inhabitants and most statistical indicators are proportional, so the sample size required for this survey was calculated in accordance with proportional errors [see Table B.2]. In consideration of the length and budget of survey, with a confidence coefficient of 95 percent, the proportional error of core indicators was less than 3.5 percent.

According to CMMR's experience with sample surveys, the design efficiency of the sampling methods used in this survey (Deff) should be kept between 1.2 and 1.3. For this survey, Deff is 1.25. As the sampling precision is under a confidence coefficient of 95 percent, and the proportional error of the core indicators was less than 3.5 percent, the minimum sample size of households in each province, n, equals to 980. So, the sample size calculation formula is as following:

n: Sample size
P: Estimated proportion of core indicator; this time, it is conservatively estimated to be 50 percent.
Z: The value of quintile under the confidence level of 95 percent is 1.96.
e: The value of anticipation error is 0.035.

Non-Household Sample Size

This time, the survey was applied to nonhousehold samples.

In every village: it was planned to visit one village leader (village head, village Communist Party of China (CPC) secretary, or village accountant, and

Table B.2 Sample Sizes for Demand Survey

Province	Questionnaire type	Design sample size	Target sample size	Number of question-naires dis-tributed	Number of question-naires recov-ered	Number of valid ques-tionnaires recovered
Jilin	Households	980	980	1,049	1,040	1,019
	Communities	80	80	80	80	80
	Infrastructure	80	80	80	80	80
	Schools	80	32	32	32	32
	Public Access Points	80	80	80	80	79
Shandong	Households	980	980	1,030	1,030	1,029
	Communities	80	80	80	79	79
	Infrastructures	80	80	80	79	79
	Schools	80	15	15	15	15
	Public Access Points	80	80	80	80	80
Guizhou	Households	980	980	1,055	1,041	1,012
	Communities	80	80	80	79	79
	Infrastructure	80	80	80	79	79
	Schools	80	65	63	63	63
	Public Access Points	80	57	50	48	48
Subtotal	Households	2,940	2,940	3,134	3,111	3,060
	Communities	240	240	240	238	238
	Infrastructure	240	240	240	238	238
	Schools	240	112	110	110	110
	Public Access Points	240	217	210	208	207
	Total	**3,900**	**3,749**	**3,934**	**3,905**	**3,853**

so on); it was planned to visit one member of staff of places that provide Internet service where villagers usually go (inside or outside the village), and it was also planned to visit one principal of the middle school and primary school in the village.

Note

1. The final version of the survey tool is available at: https://docs.google.com/open?id=0 BwCAk39OtgfRLU9tZGpKMW51SGc.

Library Survey

The questionnaires included 11 sections: (1) Services; (2) Location; (3) Human resources; (4) Users; (5) Use; (6) Internet Access; (7) Online content; (8) Information Technology (IT); (9) Projects; (10) Finance; and (11) Acquisitions. They were targeted towards different levels of the library system: Level 1 (provincial and regional libraries); Level 2 (country libraries); and Level 3 (township and village libraries).

The questionnaires were piloted in Guizhou and Shandong provinces and subsequently revised. Questionnaire distribution was administered with the help of library authorities in each province between March and May 2012. Although over 1,000 administrative villages in each of the three provinces were selected for the Level 3 questionnaire, those that actually responded varied across provinces (see table C.1).

During June and July 2012, data from the printed questionnaire returns were entered into the Survey Gizmo tool. The data were then processed and initial analyses produced between July and September 2012.

Table C.1 Response Rates for the Three Provinces

	Level 1	Level 2	Level 3
Guizhou	6	51	892
Jilin	7	24	1,262
Shandong	15	105	527

Rural ICT Impact Evaluation Surveys

Guizhou

A survey was conducted through a combination of online and printed questionnaire. First of all, the questionnaire was put on the Internet. A number of multipurpose telecenter (MT) administrators were selected across Guizhou Province, who would then recruit common users and fill in the questionnaire online together with them; for nonusers, the administrators would give them printed copies of the questionnaire and enter the answers into the online questionnaire. Two hundred and fifty-four valid copies were retrieved for this survey: 203 of them were completed by users (including administrators and common users); and 51 of them were completed by nonusers. Users included 34 administrators and 169 common users. They are from all the nine prefectures in Guizhou Province.

Jilin

A telephone interview was carried out about users' experiences with agricultural information platforms and reasons for nonuse. The sample consisted of rural residents in Jilin Province who are involved in agricultural production or sales. There were a total of 200 valid samples, including 150 users and 50 nonusers. Note that responses refer to both types of agricultural service platforms available in the province (12316, an agricultural information service similar to 12582, is provided by China Unicom).

Shandong

The study was conducted by means of telephone survey. The sample consists of users of the e-commerce platform Shandong Agricultural Development (http://www.sdxnw.gov.cn) and rural residents in Shandong, including both users and nonusers of e-commerce platforms. There are altogether 131 valid samples, including 53 users and 78 nonusers.

Information and Communications in the Chinese Countryside
http://dx.doi.org/10.1596/978-1-4648-0204-1

References

Basu, Medha. 2013. "400,000 Free Wi-Fi Hotspots in Thailand by 2014." *Asia Pacific FutureGov*, August 13. http://www.futuregov.asia/articles/2013/aug/13/400000-wi-fi-hotspots-thailand-2014/.

Batista, Jandré. 2009. "Uruguay's 'One Laptop Per Child' Policy: Small Changes in the Latin America's Education Reality." *WAVE*, April 10. http://www.wavemagazine.net/arhiva/30/science/uruguay-laptop-policy.htm.

Broadband Commission. 2013. *The State of Broadband 2013: Universalizing Broadband.* Geneva: International Telecommunication Union; Paris: UNESCO. http://www.broad-bandcommission.org/documents/bb-annualreport2013.pdf.

Cai, Jing. 2013. "Impact Evaluation Approach Report: China Rural ICT." Internal Report prepared for World Bank study on ICT use in rural China.

Chen, Li. 2003."China." In *Meta-Survey on the Use of Technologies in Education,* Bangkok: UNESCO. 2003. http://www.unescobkk.org/education/ict/online-resources/e-library/key-resources/metasurvey.

China Internet Network Information Center (CNNIC). 2007. "Survey Report on Internet Development in Rural China 2007." Beijing.

———. 2012. *Current Development of China's Internet and a Look at Internet Use in Rural Areas.* Beijing: China Internet Network Information Center.

———. 2013a. *Prospect of Business Process Outsourcing (BPO) and Microwork in Rural China.* Beijing: China Internet Network Information Center.

———. 2013b. *Evaluation of Rural ICT Interventions in Three Chinese Provinces.* Beijing: China Internet Network Information Center.

———. 2013c. "Statistical Report on Internet Development in China." Beijing.

China Mobile Limited. 2009. "Annual Report 2008." http://www.chinamobileltd.com/en/ir/reports/ar2008.pdf.

Fang Cai, John Giles, Philip O'Keefe, and Dewen Wang. 2012. *The Elderly and Old Age Support in Rural China: Challenges and Prospects.* Washington, DC: World Bank.

Gomez, Ricardo. 2012. *Libraries, Telecentres, Cybercafes and Public Access to ICT: International Comparisons.* Hershey, PA: IGI Global. http://www.igi-global.com/book/libraries-telecentres-cybercafes-public-access/49588 /.

He, Eileen, and Simon Ye. 2009. "Rural China PC Program Will Increase PC Shipments in 2009." *Gartner Research*, March 10. http://www.gartner.com/id=909330.

Liu, Chun. 2012. "Seeking the 'Best Practices' in Rural Agricultural Informatization: Evidence from China's Sichuan Province." *SSRN Scholarly Paper*, March 11. http://papers.ssrn.com/abstract=2020093.

McQuaide, Shiling. 2009. "Making Education Equitable in Rural China through Distance Learning." *The International Review of Research in Open and Distance Learning* 10 (1). http://www.irrodl.org/index.php/irrodl/article/view/590.

Mo, Di, Johan Swinnen, Linxiu Zhang, Hongmei Yi, Qinghe Qu, Matthew Boswell, and Scott Rozelle. 2013. "Can One Laptop Per Child Reduce the Digital Divide and Educational Gap? Evidence from a Randomized Experiment in Migrant Schools in Beijing." *World Development* 46 (June): 14–29. http://www.sciencedirect.com/science/article/pii/S0305750X13000077.

Montlake, Simon. 2012. "Thailand Taps China For Cut-Price 'One Tablet Per Child' Program." *Forbes*, February 22. http://www.forbes.com/sites/simonmontlake/2012/02/22/thailand-taps-china-for-cut-price-one-tablet-per-child-program/.

National Bureau of Statistics of China. 2009. *China Yearbook of Rural Household Survey 2009*. Beijing: China Statistics Press.

———. 2012. "Statistical Communiqué on the 2011 National Economic and Social Development," February 22. http://www.stats.gov.cn/english/newsandcomingevents/t20120222_402786587.htm.

Paberza, Kristine, and Ugne Rutkauskiene. 2010. "Outcomes-based Measurement of Public Access Computing in Public Libraries: A Comparative Analysis of Studies in Latvia and Lithuania." *Performance Measurement and Metrics* 11 (1): 75–82.

Pew Research Center. 2013. "Cell Phones Nearly Universal in Much of the World." Pew Research Center, February 4. http://www.pewresearch.org/daily-number/cell-phones-nearly-universal-in-much-of-the-world/.

Qiang, Christina Zhen-Wei, Asheeta Bhavnani, Nagy Hanna, Kaoru Kimura, and Randeep Sudan. 2009. *Rural Informatization in China*. Washington, DC: World Bank. https://openknowledge.worldbank.org/handle/10986/5934.

Qiu, Guanhua. 2012. "Introduction to Suzhou Central Branch Library System." Paper presented at the Workshop on China Rural ICT Project: Library Landscape Study, Beijing, December 13, 2012. http://www.worldbank.org/en/news/feature/2013/01/29/china-improving-rural-access-to-information-communication-technologies.

Rossotto, Carlo M., Siou Chew Kuek, and Cecilia Paradi-Guilford. 2012. "New Frontiers and Opportunities in Work: ICT is Dramatically Reshaping the Global Job Market." *ICT Policy Notes*. World Bank, June. http://siteresources.worldbank.org/INFORMATIONANDCOMMUNICATIONANDTECHNOLOGIES/Resources/1221302_NewFrontier_PolicyNote_LowRes.pdf.

State Information Center (SIC). 2011. "Background Report on Rural Informatization in China."

———. 2012. "Demand Survey Report of China Rural ICT."

Ting, Carol, and Famin Yi. 2012. "ICT Policy for the 'Socialist New Countryside': a Case Study of Rural Informatization in Guangdong, China." *SSRN Scholarly Paper*, April 8. http://papers.ssrn.com/abstract=2186795.

Trucano, Michael. 2010. "What Happens When 'All' Children and Teachers Have Their Own Laptops." *Edutech*. World Bank, May 14. http://blogs.worldbank.org/edutech/ceibal-archives.

———. 2012. "ICT and Rural Education in China." *Edutech*, August 24. http://blogs.worldbank.org/edutech/ict-and-rural-education-in-china.

United States Department of Agriculture (USDA Foreign Agricultural Services. 2011. *China's 12th Five-Year Plan (Agricultural Section)*. http://gain.fas.usda.gov/Recent percent20GAIN percent20Publications/China's percent2012th percent20Five-Year percent20Plan percent20(Agricultural percent20Section)_Beijing_China percent20percent20Peoples percent20Republic percent20of_5-3-2011.pdf.

Walton, M., and J. Donner. 2012. "Public Access, Private Mobile: The Interplay of Shared Access and the Mobile Internet for Teenagers in Cape Town." Global Impact Study Research Report Series. Cape Town, South Africa: University of Cape Town. http://www.globalimpactstudy.org/2012/11/mobile-internet-in-depth-study-research-report-released/.

World Bank. 2013 "China: Improving Rural Access to Information and Communication Technologies." Feature Story. http://www.worldbank.org/en/news/feature/2013/01/29/china-improving-rural-access-to-information-communication-technologies.

World Bank and the Development Research Center of the State Council, P. R. China. 2013. *China 2030: Building a Modern, Harmonious, and Creative Society*. Washington, DC: World Bank. http://www.worldbank.org/content/dam/Worldbank/document/China-2030-complete.pdf.

Wu, Xiao. 2012. "The Road to Digital Resources Sharing: Cases of the Cultural Information Resources Management Center of China." Paper presented at IFLA World Library and Information Congress, Helsinki, August 11–17. http://conference.ifla.org/past/2012/79-wu-en.pdf.

Xinhua. 2011. "Guizhou Started Construction of Village-level Multi-functional Information Service Station." Xinhua, May 31. http://www.sourcejuice.com/1448405/2011/05/31/Guizhou-started-construction-village-level-multi-functional-information-service/.

Yongtao, Shen. 2009. "The Rural ICT Development in China." Paper presented at the Regional Workshop on Community e-Centres for Rural Development, New Delhi, October 29.

Yu, Liangzhi. 2010. "Information Worlds of Chinese Farmers and Their Implications for Agricultural Information Services: a Fresh Look at Ways to Deliver Effective Services." Paper presented at IFLA World Library and Information Congress, Gothenburg, Sweden, August 10–15. http://conference.ifla.org/past/ifla76/216-2.htm.

Yu, Ying, and Xiangyang Qin. 2011. "Study on Rural Cultural and Entertainment Informatization Status in China." In *2011 International Conference on Future Information Technology*. Singapore: IACSIT Press. http://ipcsit.com/list-50-1.html.

Zeng, Haijun, Ronghuai Huang, Yuchi Zhao, and Jinbao Zhang. 2012. "ICT and ODL in Education for Rural Development: Current Situation and Good Practices in China." International Research and Training Centre for Rural Education (INRULED), UNESCO. http://www.inruled.org/iERD/Publication/iERD%20in%20China%20for%20eLA%20(UNESCO-INRULED).pdf.

Zhang, Guangqin, and Robert Davies. 2013. "China Library Landscape Study." https://docs.google.com/file/d/0BwCAk39OtgfRSmNfcXk0bjBidm8/edit?usp=sharing.

Zhang, Leilei, and Jinmin Hao. 2012. "Librarians 2.0: IT Literacy of Librarians in China." Paper presented at the IFLA World Library and Information Congress, Helsinki, August 11–17. http://conference.ifla.org/past/ifla78/session-150.htm.

Zhang, Linxlu, Fang Lai, Yaojiang Shi, Matthew Boswell, and Scott Rozelle. 2013. "The Roots of Tomorrow's Digital Divide: Documenting Computer Use and Internet Access in China's Elementary Schools Today." *China & World Economy* 21 (3): 61–79. http://onlinelibrary.wiley.com/doi/10.1111/j.1749-124X.2013.12022.x/abstract.

Zhou, Hongren. 2007. "E-Government Funding in China." In *The 7th Global Forum on Reinventing Government*. Vienna, Austria.

Environmental Benefits Statement

The World Bank is committed to reducing its environmental footprint. In support of this commitment, the Publishing and Knowledge Division leverages electronic publishing options and print-on-demand technology, which is located in regional hubs worldwide. Together, these initiatives enable print runs to be lowered and shipping distances decreased, resulting in reduced paper consumption, chemical use, greenhouse gas emissions, and waste.

The Publishing and Knowledge Division follows the recommended standards for paper use set by the Green Press Initiative. Whenever possible, books are printed on 50 percent to 100 percent postconsumer recycled paper, and at least 50 percent of the fiber in our book paper is either unbleached or bleached using Totally Chlorine Free (TCF), Processed Chlorine Free (PCF), or Enhanced Elemental Chlorine Free (EECF) processes.

More information about the Bank's environmental philosophy can be found at http://crinfo.worldbank.org/wbcrinfo/node/4.

green
press
INITIATIVE

www.ingramcontent.com/pod-product-compliance
Lightning Source LLC
Chambersburg PA
CBHW080001280326
41935CB00013B/1719